Diesel Engines and Electric Power

Unit I, Lesson 8
Third Edition

by Ron Baker

Published by

PETROLEUM EXTENSION SERVICE
THE UNIVERSITY OF TEXAS AT AUSTIN
Continuing & Innovative Education
Austin, Texas

in cooperation with

INTERNATIONAL ASSOCIATION
OF DRILLING CONTRACTORS
Houston, Texas

1998

Library of Congress Cataloging-in-Publication Data

Baker, Ron, 1940—

    Diesel engines and electric power / by Ron Baker. — 3rd ed.

      p. cm. — (Rotary drilling series ; unit I, lesson 8)

    "In cooperation with International Association of Drilling

Contractors, Houston, Texas."

    ISBN 0-88698-169-7 (alk. paper)

    1. Oil well drilling, Electric—Power supply. 2. Oil well
      drilling rigs—Power supply. 3. Diesel motor. 4. Electric
      power.
      I. International Association of Drilling Contractors. II. Title.

III. Series.

TN871.2.B3157   1998

622'.3381—dc21   98-18155

                                              CIP

Catalog no. 2.108301
ISBN 0-88698-169-7

Contents

Figures

Foreword

For many years, the Rotary Drilling Series has oriented new personnel and further assisted experienced hands in the rotary drilling industry. As the industry changes, so must the manuals in this series reflect those changes.

The revisions to both text and illustrations are extensive. In addition, the layout has been "modernized" to make the information easy to get; the study questions have been rewritten; and each major section has been summarized to provide a handy comprehension check for the student.

PETEX wishes to thank industry reviewers—and our readers—for invaluable assistance in the revision of the Rotary Drilling Series. On the PETEX staff, Deborah Caples designed the layout; Doris Dickey proofread innumerable versions; and Kathryn Roberts saw production through from idea to book.

Although every effort was made to ensure accuracy, this manual is intended to be only a training aid; thus, nothing in it should be construed as approval or disapproval of any specific product or practice.

Ron Baker

Acknowledgments

Special thanks to Ken Fischer, Vice-President of Member Services, International Association of Drilling Contractors, who reviewed this manual and secured other reviewers. Tom Thomas and Michel Moy of Sedco Forex Schlumberger provided invaluable suggestions on the content and language. Without their assistance, this book could not have been written. In addition, special thanks to Leslie Kell, who managed to interpret some difficult sketches and make them into excellent drawings, and to Brandt/Tuboscope, Derrick Equipment Corp., M-D Totco, Mission Fluid King, and National Oilwell for assistance in obtaining photos.

In spite of the assistance PETEX received in writing and illustrating this book, PETEX is solely responsible for its contents. Also, while every effort was made to ensure accuracy, this manual is intended only as a training aid. Nothing in it should be considered approval or disapproval of any specific product or practice.

Units of Measurement

Throughout the world, two systems of measurement dominate: the English system and the metric system. Today, the United States is almost the only country that employs the English system.

The English system uses the pound as the unit of weight, the foot as the unit of length, and the gallon as the unit of capacity. In the English system, for example, 1 foot equals 12 inches, 1 yard equals 36 inches, and 1 mile equals 5,280 feet or 1,760 yards.

The metric system uses the gram as the unit of weight, the metre as the unit of length, and the litre as the unit of capacity. In the metric system, for example, 1 metre equals 10 decimetres, 100 centimetres, or 1,000 millimetres. A kilometre equals 1,000 metres. The metric system, unlike the English system, uses a base of 10; thus, it is easy to convert from one unit to another. To convert from one unit to another in the English system, you must memorize or look up the values.

In the late 1970s, the Eleventh General Conference on Weights and Measures described and adopted the Système International (SI) d'Unités. Conference participants based the SI system on the metric system and designed it as an international standard of measurement.

The *Rotary Drilling Series* gives both English and SI units. And because the SI system employs the British spelling of many of the terms, the book follows those spelling rules as well. The unit of length, for example, is *metre*, not *meter*. (Note, however, that the unit of weight is *gram*, not *gramme*.)

To aid U.S. readers in making and understanding the conversion to the SI system, we include the following table.

English-Units-to-SI-Units Conversion Factors

Quantity or Property	English Units	Multiply English Units By	To Obtain These SI Units
Length, depth, or height	inches (in.)	25.4	millimetres (mm)
		2.54	centimetres (cm)
	feet (ft)	0.3048	metres (m)
	yards (yd)	0.9144	metres (m)
	miles (mi)	1609.344	metres (m)
		1.61	kilometres (km)
Hole and pipe diameters, bit size	inches (in.)	25.4	millimetres (mm)
Drilling rate	feet per hour (ft/h)	0.3048	metres per hour (m/h)
Weight on bit	pounds (lb)	0.445	decanewtons (dN)
Nozzle size	32nds of an inch	0.8	millimetres (mm)
Volume	barrels (bbl)	0.159	cubic metres (m³)
		159	litres (L)
	gallons per stroke (gal/stroke)	0.00379	cubic metres per stroke (m³/stroke)
	ounces (oz)	29.57	millilitres (mL)
	cubic inches (in.³)	16.387	cubic centimetres (cm³)
	cubic feet (ft³)	28.3169	litres (L)
		0.0283	cubic metres (m³)
	quarts (qt)	0.9464	litres (L)
	gallons (gal)	3.7854	litres (L)
	gallons (gal)	0.00379	cubic metres (m³)
	pounds per barrel (lb/bbl)	2.895	kilograms per cubic metre (kg/m³)
	barrels per ton (bbl/tn)	0.175	cubic metres per tonne (m³/t)
Pump output and flow rate	gallons per minute (gpm)	0.00379	cubic metres per minute (m³/min)
	gallons per hour (gph)	0.00379	cubic metres per hour (m³/h)
	barrels per stroke (bbl/stroke)	0.159	cubic metres per stroke (m³/stroke)
	barrels per minute (bbl/min)	0.159	cubic metres per minute (m³/min)
Pressure	pounds per square inch (psi)	6.895	kilopascals (kPa)
		0.006895	megapascals (MPa)
Temperature	degrees Fahrenheit (°F)	$\dfrac{°F - 32}{1.8}$	degrees Celsius (°C)
Thermal gradient	1°F per 60 feet	—	1°C per 33 metres
Mass (weight)	ounces (oz)	28.35	grams (g)
	pounds (lb)	453.59	grams (g)
		0.4536	kilograms (kg)
	tons (tn)	0.9072	tonnes (t)
	pounds per foot (lb/ft)	1.488	kilograms per metre (kg/m)
Mud weight	pounds per gallon (ppg)	119.82	kilograms per cubic metre (kg/m³)
	pounds per cubic foot (lb/ft³)	16.0	kilograms per cubic metre (kg/m³)
Pressure gradient	pounds per square inch per foot (psi/ft)	22.621	kilopascals per metre (kPa/m)
Funnel viscosity	seconds per quart (s/qt)	1.057	seconds per litre (s/L)
Yield point	pounds per 100 square feet (lb/100 ft²)	0.48	pascals (Pa)
Gel strength	pounds per 100 square feet (lb/100 ft²)	0.48	pascals (Pa)
Filter cake thickness	32nds of an inch	0.8	millimetres (mm)
Power	horsepower (hp)	0.75	kilowatts (kW)
Area	square inches (in.²)	6.45	square centimetres (cm²)
	square feet (ft²)	0.0929	square metres (m²)
	square yards (yd²)	0.8361	square metres (m²)
	square miles (mi²)	2.59	square kilometres (km²)
	acre (ac)	0.40	hectare (ha)
Drilling line wear	ton-miles (tn•mi)	14.317	megajoules (MJ)
		1.459	tonne-kilometres (t•km)
Torque	foot-pounds (ft•lb)	1.3558	newton metres (N•m)

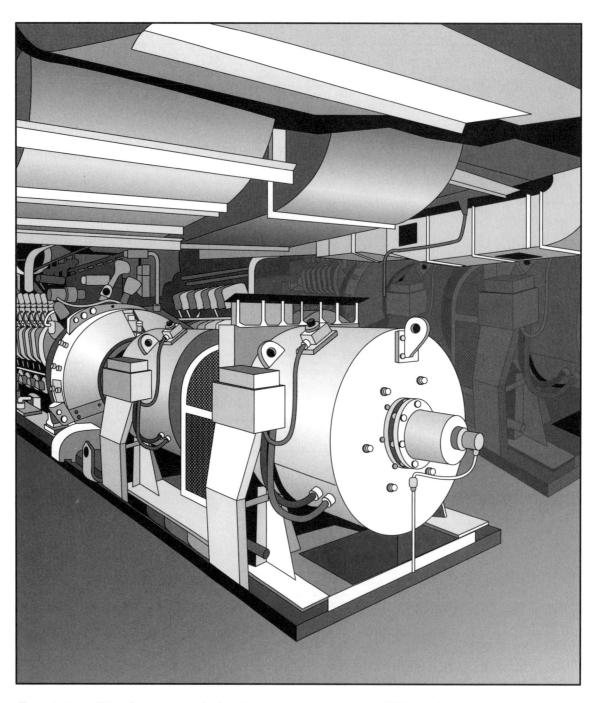

Frontispiece: Diesel engines and electric generators power an offshore rig.

DIESEL ENGINES
Introduction

▼
▼
▼

he main purpose of a rotary rig is to drill, or make, a hole. To make hole, the rig must have a source of power. What is more, the rig must be able to transmit this power to equipment that needs it. For example, the mud pumps need power to move drilling fluid. The drawworks also needs power to do its work.

Usually, large internal combustion engines power the rig (fig. 1). A mixture of fuel and air burns inside the engine to make it run. If the engine is running correctly, the fuel-air mixture burns at a controlled rate. Keep in mind that an engine must get oxygen from the atmosphere before the fuel can burn.

Engine Power and Transmission

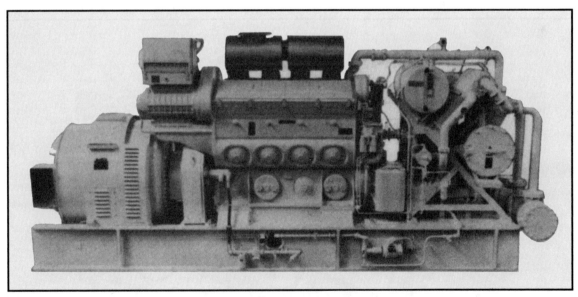

Figure 1. Large internal combustion engines power drilling rigs.

Power Transmission

A rig owner does not install an engine on each piece of equipment that needs power. It is more practical to set up two, three, or more powerful engines at a single place on the site. Special equipment attached to the engines then transmits power to the equipment that needs it. Offshore, for example, the rig builder may put three or four engines in an engine room that is some distance from the drawworks and the mud pumps. On land rigs, the rig-up crew often places the engines next to the rig floor, or crew members may place them inside a special house or shed that is several yards (metres) from the equipment needing the power.

Mechanical Transmission and the Compound

Some rigs use machinery (gears, sprockets, and chains) to transmit engine power. If machinery transmits power, the rig has a mechanical transmission. In a mechanical transmission, the rig builders mount couplings on each engine. These couplings connect the engines to the machinery that transmits the engine power.

Crew members call the heavy-duty sprockets and chains that make up this machinery the compound. The rig-up crew connects the compound to the engine couplings and to the equipment needing power, such as the drawworks and the mud pumps. The compound can then transfer engine power to the equipment (fig. 2).

Figure 2. Prime movers and the compound in a mechanical drive

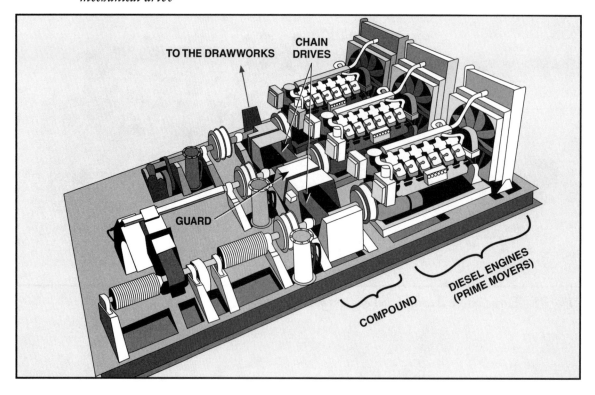

A steel shroud (guard) surrounding the compound safeguards personnel. It also keeps in a spray lubricant that oils the moving parts. (For more information about compounds, see Unit I, Lesson 6 in the IADC–PETEX Rotary Drilling Series, entitled The Drawworks and the Compound.)

Many rigs do not use a compound to transmit engine power. Instead, they use electricity. On electric rigs, a large electric generator is attached to each engine. The engines run the generators (fig. 3). The generators, in turn, make electricity, which they send through heavy-duty wires (cables) and special controls to powerful electric motors (fig. 4). These electric motors power the equipment. Usually, the rig owner places the motors on or next to the equipment being powered.

Electric Transmission

Figure 3. Diesel engine and generator in an electric drive system

Figure 4. Powerful electric motors drive a mud pump.

3

Engines Versus Motors

An engine and a motor are similar devices: both provide power to drive the equipment. They are so similar that many people call an engine a motor. Strictly speaking, however, an engine and a motor are different. An engine changes thermal energy into power to produce force and motion. A motor, on the other hand, creates power without having to change or transform its energy source. Energy is neither created nor destroyed, although its form may be changed.

On rigs, for example, the engines change the heat energy from burning fuel into mechanical energy. The mechanical energy runs generators to make electrical energy. Using this electrical energy, the motors then power equipment; however, the motors run on electricity and therefore do not convert it to another form of energy.

Diesel Engines

Engines take in air and fuel. They burn this air and fuel mixture to create energy to do work.

    Engines burn fuel and air to move pistons up and down in cylinders (fig. 5). Intake valves or intake ports let air into each cylinder, where it mixes with the fuel. A spark or some other heat source ignites the mixture of air and fuel. As the fuel-air mixture burns, it expands to move the pistons. Burned gases leave the cylinders through exhaust valves.

How Engines Operate

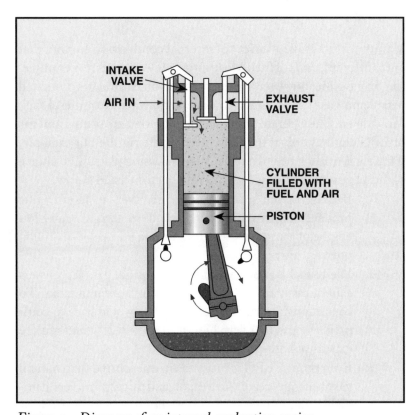

Figure 5. Diagram of an internal combustion engine

Pistons are attached to connecting rods, also known as piston rods. The connecting rods are attached to a solid heavy-duty steel shaft called the crankshaft. Heavy-duty rod bearings ride in the connecting rods where they fit on the crankshaft. Heavy-duty bearings are also fitted on the crankshaft where it rotates in the engine crankcase, the main body of the engine. These are called main bearings. The engine's lubrication system supplies oil to these bearings, reducing friction between the moving parts.

The connecting rods transfer the piston's up-and-down (reciprocating) motion to the crankshaft, causing it to rotate. Connected to the rotating crankshaft is a heavy metal disk called a flywheel. The heavy flywheel's momentum gives an almost uniform rotational speed to the crankshaft and to the connected machinery.

To operate, rigs take the engine's power from the flywheel. As stated earlier, on many rigs, the rotating flywheel turns an electric generator to make electricity. The electricity runs motors, which power the rig's equipment. On other rigs, the flywheel turns a special coupling that is connected to the compound, which in turn transfers engine power to the rig equipment.

Engine Fuels

Manufacturers make a variety of internal combustion engines that burn different kinds of fuel. Most automobile engines, for example, run on gasoline; however, rig engines run on natural gas, liquefied petroleum gas (LPG) such as butane and propane, or diesel fuel. Most rig engines operate on diesel fuel. Rig owners use natural gas or LPG engines only if the fuels are near the rig site, for example, if the rig is drilling near a natural gas processing plant. The choice of diesel power for rigs is based on the factors listed below.

1. Because of the special way it uses fuel, a diesel engine produces more twisting force (torque) than a gas or an LPG engine of the same size. More torque means more drilling power.

2. Diesel fuel is more portable than natural gas. Rig owners cannot easily transport natural gas or store it in tanks. To have enough to run the engines, they would have to compress the gas to a very high degree, which would require expensive high-pressure tanks.

3. Even though LPG is more easily transported than natural gas, it changes readily to a vapor, and this vapor is very flammable. Because diesel fuel does not vaporize as readily as LPG, it is safer to transport, handle, and store than is LPG.

Natural gas, LPG, and diesel engines differ in two main ways—
1. they use different fuels, and
2. they use different methods to ignite the fuels.

Gas or LPG Engines and Diesel Engines

Natural gas and LPG engines use spark plugs. A spark plug creates a high-voltage electric spark that fires, or ignites, the fuel-air mixture inside the engine's cylinders. Gas and LPG engines are therefore spark ignition (SI) engines.

Spark Ignition

Some diesel engines have glow plugs. Although these may look like spark plugs, glow plugs only help start a diesel engine in cold weather by preheating the combustion chamber. They do not ignite the fuel-air mixture after the engine starts. In the text that follows, you will see how diesels ignite fuel without spark plugs.

Combustion is the controlled burning of fuel and air. Natural gas and LPG (SI) engines draw fuel and air into the cylinders, where combustion occurs. To run well, SI engines need the right proportion, or ratio, of fuel to air. While a mixture of too much fuel with too little air may not burn, mixing too little fuel with too much air may also fail to produce combustion.

Fuel-to-Air Ratio in Spark Ignition (SI) Engines

Gas and LPG engines run best on a mixture of about 15 parts air to 1 part fuel. This 15-to-1 ratio holds true no matter how little or how much the engine throttle is opened. So when the driller opens the throttle to increase the engine's speed, more fuel goes into the cylinders. At the same time, more air also goes in to keep the mixture at about 15 to one.

Conversely, when the driller decreases the engine's speed, less fuel and less air go into the cylinders. Designers of gas and LPG engines build them so that the fuel-to-air ratio stays at about 15 to 1 regardless of speed.

As mentioned before, spark plugs do not ignite the fuel-air mixture in a diesel engine; very hot air ignites the fuel. Anytime you compress air, its temperature increases. Compress it enough, and the temperature becomes hot enough to ignite fuel.

Compression Ignition (CI) Engines

Engine pistons compress the air inside the cylinders. In a diesel, they compress the air so much that the fuel-air mixture ignites. Diesel engines are therefore compression ignition (CI) engines. A diesel's piston compresses the air until the temperature reaches about 1,000° Fahrenheit (F) or about 540° Celsius (C). A fuel injector then sprays diesel fuel into the cylinder, where it ignites.

Fuel-to-Air Ratio in Compression Ignition (CI) Engines

Think about a spark-ignition engine for a moment. Suppose the driller needs to speed the engine up. To do so, the driller moves the engine's throttle to increase the amount of fuel going into the cylinders. At the same time, the engine also takes in more air. Remember that the fuel-to-air ratio has to be about 15 to 1 in an SI engine, regardless of its speed.

Now consider a diesel engine. Diesels (CI engines) work differently from SI engines. CI engines draw in a constant amount of air, regardless of their speed. When the driller moves the throttle to increase the fuel to the cylinders, the engine does not take in any more air. Since the air in the cylinder is so hot, diesel fuel and air ignite no matter how much fuel the injector puts in. The fuel-to-air ratio is not as critical in a CI engine, as long as enough oxygen exists to support combustion.

Strokes and Cycles

Two types of diesel engines are four-strokes-per-cycle and two-strokes-per-cycle. A cycle is a series of events that happen when an engine runs. A stroke is the piston's upward-or-downward movement in the engine's cylinders. Strokes-per-cycle is the number of strokes a piston makes to complete one operating cycle.

Two-strokes-per-cycle means that the piston goes down one stroke and up one stroke to complete one operating cycle. Four-strokes-per-cycle means that the piston goes down one stroke, up a second stroke, down again for a third stroke, and then up again for a fourth stroke to complete one operating cycle.

A piston makes one stroke downward when it goes from its highest point in the cylinder to its lowest point. It makes one stroke upward when it goes from its lowest point in the cylinder to its highest point (fig. 6). The highest point is top dead center, or TDC. The lowest point is bottom dead center, or BDC.

Most people use a shorter expression for two-strokes-per-cycle and four-strokes-per-cycle. Some shorten the terms to two-cycle and four-cycle, while others use the expressions two-stroke and four-stroke. The latter expressions will be used in this manual.

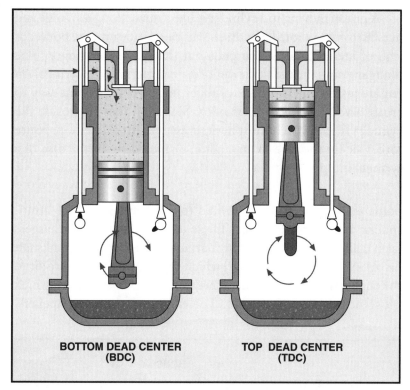

BOTTOM DEAD CENTER
(BDC)

TOP DEAD CENTER
(TDC)

Figure 6. At left, piston is at bottom dead center (BDC); at right, piston is at top dead center (TDC).

Manufacturers get air into a diesel engine's cylinders in one of two ways—

1. they naturally aspirate it; or
2. they force air into it.

Aspiration is the drawing in of air by suction. Natural aspiration means that the engine draws in air without aid from any special devices. A naturally aspirated engine, therefore, draws air in without assistance as it runs.

Pistons draw air in as they move down the cylinders. When a piston moves down, it creates an area of reduced pressure—a vacuum—above it. Since the piston creates a vacuum, air from the atmosphere rushes in to fill the vacuum. In other words, the engine draws in air that is at atmospheric pressure. Atmospheric pressure at sea level is about 14.7 pounds per square inch (psi) or about 100 kilopascals (kPa).

Forced-Air Induction and Natural Aspiration

Natural Aspiration

A piston moving in a cylinder is like a nurse using a syringe and needle to give a shot. The nurse sticks the needle into the bottle, or vial, of medicine and then pulls out the syringe's plunger. The plunger draws the medicine out of its vial and into the syringe. An engine piston and a syringe plunger both work the same way to create a vacuum inside a cylinder. Since the pressure inside the cylinder is less than atmospheric pressure, atmospheric pressure forces air (in the case of an engine) or medicine (in the case of a syringe) into the cylinder.

Forced-Air Induction   Some engines use special devices to compress and force atmospheric air into the cylinder. These devices can be superchargers (also called blowers) or turbochargers. The engine's crankshaft drives a supercharger (blower) which supplies high-pressure air to the engine (fig. 7). Inside the blower are blades or impellers, which act something like fan blades. The engine turns these impellers.

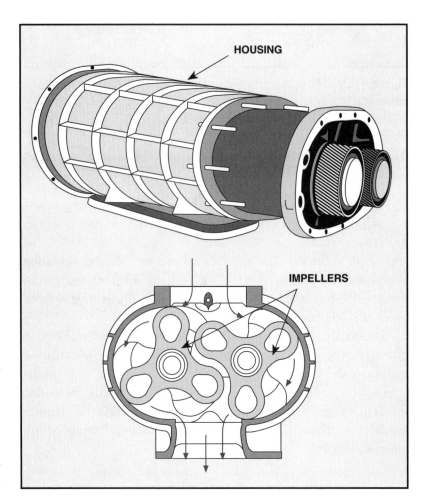

Figure 7. Supercharger (blower)

Since a case or housing encloses the impellers, they raise the surrounding air pressure above atmospheric level. In other words, the impellers compress air and force it into the engine. At the right moment, the engine cylinders take in this compressed air.

Some engines have turbochargers that, like blowers, force air into the cylinders. Unlike blowers, however, which the engine crankshaft drives, turbochargers are driven by engine exhaust gases.

Forcing air into an engine makes it more powerful. Since the supercharger or turbocharger compresses the air within it, that air is denser (weighs more) than atmospheric air. Forced-air induction thus packs more air into the cylinder. With more air, the fuel can burn more completely, which results in more power. Typically, a supercharged or turbocharged diesel engine is about one-third more powerful than a naturally aspirated engine of the same size.

Forced-Air Induction and Power

Manufacturers can supercharge or turbocharge both two-stroke and four-stroke diesel engines. Only four-stroke diesel engines, however, can be naturally aspirated. You will learn why two-stroke diesels must be supercharged or turbocharged in the section on two-stroke diesel engines. (Note that both two-stroke and four-stroke gasoline engines can be supercharged, turbocharged, or naturally aspirated.)

To summarize—

How a rig transmits power

- If the rig is mechanical, special couplings on the engine transfer engine power to sprockets and chains called the compound. The compound transmits engine power to the rig equipment, such as the drawworks and mud pumps.
- If the rig is electrical, the engines turn generators that make electricity, which goes to electric motors. The electric motors, which are mounted on or near the equipment needing power, provide power to the equipment.

How engines work

- A burning fuel-and-air mixture powers pistons, which move up and down (reciprocate) in cylinders.
- The moving pistons turn a crankshaft, which, in turn, moves a flywheel to create mechanical energy.
- The mechanical energy powers a generator or a compound.

Engine fuels

- Most rig engines run on diesel fuel, although a few may run on natural gas or LPG.
- Diesel engines are compression ignition (CI) engines.
- Gas and LPG engines are spark ignition (SI) engines.

Strokes and cycles

- Diesel engines are either four-stroke-per-cycle or two-stroke-per-cycle engines.
- Two strokes per cycle means that the piston goes down one stroke and up one stroke to complete a firing cycle.
- Four strokes per cycle means that the piston goes down one stroke, up a second stroke, down again for a third stroke, and up again for the fourth stroke.

Forced-air induction and natural aspiration

- Natural aspiration means the engine draws in air from the atmosphere without any mechanical (or other) assistance.
- Forced-air induction means that a machine (a supercharger or a turbocharger) supplies air to the engine at a pressure higher than that of the atmosphere.
- Supercharged and turbocharged engines are more powerful than naturally aspirated engines of the same size because the compressing atmospheric air packs more oxygen into the cylinder. With more oxygen, the fuel can burn more completely, which results in more power.

Figure 8 shows the four strokes of the firing cycle of a naturally aspirated, four-stroke diesel engine. Note the piston, the cylinder, the intake valve, and the exhaust valve.

Four-Stroke Diesel Engines

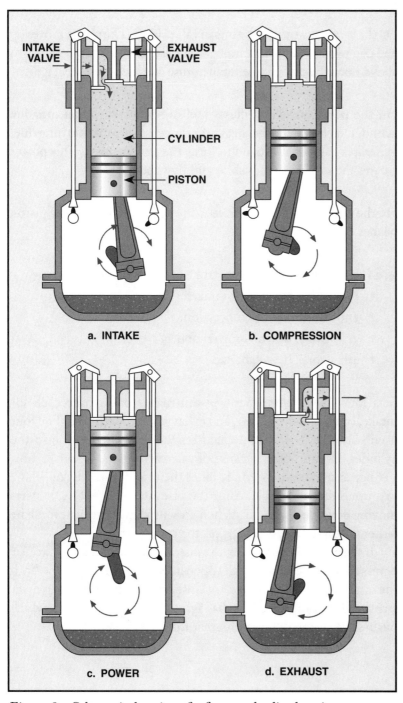

Figure 8. Schematic drawing of a four-stroke diesel engine

Intake Stroke

On the *intake stroke* of the cycle, the piston moves down. This downward movement creates a vacuum in the cylinder. Atmospheric pressure is now higher than the pressure in the cylinder, causing air to enter through the open *intake valve* and fill the vacuum.

Compression Stroke

On the *compression stroke*, the intake valve closes. The piston moves up and compresses the air. Compression raises the air's temperature to about 1,000°F (540°C), the temperature at which diesel fuel ignites.

Power Stroke

On the power stroke, a diesel fuel injector sprays fuel into the cylinder. Ignited by the heat of compression, the burning fuel generates gases that expand to force the piston down. This power rotates the engine's crankshaft and flywheel.

Exhaust Stroke

On the *exhaust stroke*, the *exhaust valve* opens, and the rising piston pushes out the used gases.

The Four-Stroke Cycle

In a four-stroke engine, four strokes make up a firing cycle—
- 1st stroke, down—air intake;
- 2nd stroke, up—compression;
- 3rd stroke, down—power; and
- 4th stroke, up—exhaust.

Four-Stroke Engines and Valve Action

In a four-stroke engine, the piston makes one stroke each for intake, compression, power, and exhaust to occur (fig. 8). The four events do not, however, depend on the piston's position in the cylinder. Events occur because of valve action.

For example, the piston reaches the top of its travel, or TDC, two times during a cycle. After the piston reaches TDC, it starts moving down the cylinder. When moving down, the piston can be either on the intake stroke or on the power stroke.

If the intake valve is open, the piston is on the intake stroke. Air enters the cylinder through the open intake valve. If, however, both the intake and exhaust valves are closed, the piston is on the power stroke. Hot compressed air has ignited the injected fuel, which pushes the piston down with great force.

Let's assume that when the cycle begins, the piston is moving up the cylinder to rotate the crankshaft. Keep in mind that the crankshaft rotates 360 degrees to make one complete revolution.

Four-Stroke Firing Details

Intake Stroke

As 25 degrees of crankshaft travel before the piston reaches TDC, the intake valve begins to open. Because the valve needs time to open enough to let air in, the valve begins to open before the piston reaches TDC.

If the intake valve did not begin opening until the piston reached TDC, the piston would already be going down by the time air got in. Such timing would reduce the volume of intake air, which would reduce the engine's power. Opening the intake valve before the piston reaches TDC ensures that air enters as the piston starts down.

As the piston moves down the cylinder, it creates a vacuum that draws air through the open intake valve. Thus, on the intake stroke, the intake valve is open, the exhaust valve is closed, and the piston is moving down the cylinder.

When the piston reaches the bottom of its travel (BDC), air continues to fill the cylinder through the still-open intake valve. The intake valve remains open for a short time after the piston reaches BDC and starts back up on the compression stroke.

Compression Stroke and Ignition

The compression stroke begins as the intake valve closes, when the crankshaft has rotated about 35 degrees after BDC. As the piston moves up the cylinder, it compresses the air trapped inside the closed cylinder. For each psi (7 kPa) increase in pressure, the air temperature rises about 2°F (1°C).

Assume that the pressure is 14.7 psi (100 kPa) and the temperature 80°F (27°C) at the beginning of the compression stroke. The piston must compress the air to 460 psi (3,500 kPa) for the air temperature to reach 1,000°F (540°C). At this pressure and temperature, diesel fuel ignites.

Ignition Stroke

Near the top of the compression stroke, both the intake and the exhaust valves close. An instant before the piston reaches TDC, the injector sprays fuel into the cylinder. The fuel enters the combustion chamber, spreads out, and starts to ignite just before the piston reaches TDC. Because the fuel starts to burn just before the piston reaches TDC, the piston is able to use the burning fuel and expanding gases as it passes TDC and travels down the cylinder.

If the injector injected fuel at the same time the piston reached TDC, the piston would already be moving down the cylinder by the time ignition occurred, resulting in a considerable loss of power.

When properly timed, the injectors put fuel into the cylinder at about 10 degrees of crankshaft travel before TDC. At 10 degrees before TDC, the piston is only about 0.005 to 0.010 of an inch (in.) or 0.127 to 0.254 of a millimetre (mm) from its topmost position. The amount the crankshaft moves when the piston is near TDC or BDC is quite large when compared to the amount the piston moves.

Power Stroke

On the power stroke, the fuel burns as expanding gases force the piston down. The power stroke shows a major difference between a diesel engine and a spark ignition engine. Because an SI engine draws in fuel with intake air, compression must be low to keep the temperature low and prevent the fuel-air mixture from igniting.

When the piston reaches TDC in an SI engine, a spark plug ignites the fuel-air mixture. This mixture burns very rapidly, causing an instantaneous push on the piston. Since this mixture burns fast, the power stroke does not last long, especially when compared to a diesel engine.

Diesel Fuel Injection

One advantage of a diesel over an SI engine is that a diesel injects a large amount of fuel. Also, diesel fuel burns more slowly, over a longer period of time, than gas or LPG. Ignition of this large amount of relatively slow-burning fuel causes a high-pressure rise on top of the piston. When this slug of fuel burns, the piston receives a very large push to begin the long power stroke.

As the piston moves down the cylinder on the power stroke, the injectors put more fuel into the already expanding gases on top of the piston. This additional fuel burns, expands, and causes a continuing rise in pressure on top of the piston. As a result, a diesel engine has a great deal of lugging power, because the fuel burns continuously over a longer period compared to gas or LPG.

Lugging power relates to the twisting force (torque) delivered to the flywheel. A diesel engine produces more torque than an SI engine of the same size. The reasons are as follows:

1. Diesel engine injectors continue to spray fuel into the cylinder during the power stroke. This additional fuel burns to create great force, and thus more torque, on top of the piston. An SI engine, on the other hand, takes in all its fuel and air during the intake stroke. During the power stroke, an SI engine does not get more fuel to create additional torque.

2. The power stroke of a diesel engine is longer than the power stroke of an SI engine. The burning gases on top of a diesel cylinder, therefore, have more time to expand and create torque.

A diesel's lugging power makes it a good choice for drilling rigs. Rig engines must produce a lot of power under a heavy load.

Exhaust Stroke

The power stroke ends while the piston is still moving down the cylinder. At about 30 degrees before BDC, the exhaust valve opens and the exhaust stroke begins. The exhaust valve opens before BDC to give the piston plenty of time to remove the burned exhaust gases from the cylinder.

The piston passes BDC and moves up to force the exhaust gases out the open exhaust valve. The exhaust valve remains open until about 20 degrees after the piston passes TDC. Then the intake stroke begins again. In the meantime, the intake valve opened before the piston reached TDC on the exhaust stroke. Both the intake and exhaust valves being open at the same time is *valve overlap.*

Valve overlap allows the exhaust gases to escape completely, which ensures a clean cylinder. Overlap also allows inlet air to fill the cylinder completely by the time the piston reaches BDC. Moreover, valve overlap lets some of the cool incoming air mix with the escaping hot exhaust gases. The incoming air cools the exhaust valve to extend its life.

Lugging Power

Cutaway of a Four-Stroke Diesel Engine

A cross section of a four-strokes-per-cycle diesel engine (fig. 9) shows many of its parts.

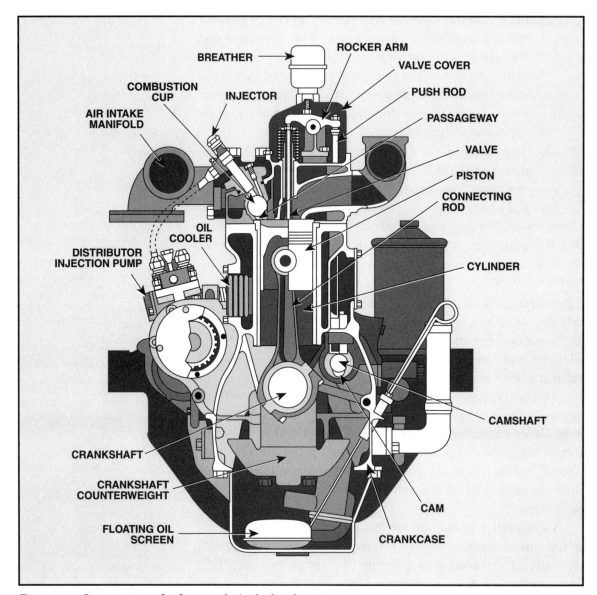

Figure 9. Cross section of a four-stroke/cycle diesel engine

On top of the valve cover in figure 9 is a *breather* (also termed a *breather cap*). Air, as it expands or contracts with changes in temperature, moves in or out of the valve cover and crankcase through the breather. The crankcase is the lower body of the engine, which houses the crankshaft. The valve cover is a metal shroud bolted over the valves that protects the valves and keeps lubricating oil from escaping. The breather also vents blow-by. *Blow-by* is high-pressure combustion gas that escapes between the piston and cylinder. By venting blow-by, the breather prevents a buildup of crankcase pressure.

Breather Cap and PCV Valve

Many engines have *positive crankcase ventilation* (PCV) *valves* (not shown in the illustration) installed between the breather and the air-intake manifold. A manifold is a series of connected pipes through which substances such as air, water, or other fluids flow. Usually, a manifold serves to join several pipes into one main pipe. The *air-intake manifold* is the piping through which the engine draws intake air into the cylinders. PCV valves direct blow-by to the intake manifold, where the pistons draw it into the cylinders for recycling. PCV valves cut down on air pollution. They also remove corrosive fumes from the crankcase and prevent sludge formation.

Some engines have a *combustion cup* in the combustion chamber (see fig. 9). The precombustion chamber houses the injector. Not all diesels have combustion cups, but on those that do, the air heated by compression accumulates in this chamber. When the injector shoots diesel fuel into the chamber, ignition takes place. The expanding gases travel down the passage from the cup to the top of the piston.

Combustion Cup

The illustrated engine's *injector* is a spray *nozzle* that breaks diesel fuel into a large number of tiny droplets in a process called *atomizing*. Air surrounds these droplets so they can easily ignite. A spray nozzle contains a spring-loaded *plunger*, which is a steel rod, or shaft, that moves up and down in the injector to supply fuel to the spray nozzle.

Fuel Injector

Many devices are available for delivering fuel to an engine's injectors. In figure 9, a *distributor* injection pump delivers fuel to each injector. This fuel distributor is similar to the electrical distributor on an automobile engine. Instead of sending electricity to several spark plugs, however, the fuel distributor uses precision timing to send fuel to several injectors.

To summarize—

Four-stroke firing cycle

- 1st stroke, down—air intake
- 2nd stroke, up—compression
- 3rd stroke, down—power
- 4th stroke, up—exhaust
- On the compression stroke, the piston compresses the air in the cylinder so much that it reaches 1,000°F (540°C).
- On the ignition stroke, the fuel injector sprays fuel into the cylinder, where it ignites because of the very hot air.

Two-Stroke Diesel Engines

A *two-stroke diesel engine* must have a supercharger or a turbocharger. These devices force air into the cylinder. At the same time, they also remove exhaust gases. Figure 10 shows the intake, compression, power, and exhaust strokes of a two-strokes-per-cycle diesel. On the intake stroke, the supercharger (blower) forces air into the cylinder through ports around the cylinder wall when the piston is at BDC. Simultaneously, the blower removes (*scavenges*) exhaust gases through the open exhaust valves.

On the compression stroke, the piston moves up, the exhaust valves close, and the piston closes the intake ports. The piston then compresses the trapped air. The injector sprays fuel into the cylinder just before the piston reaches TDC. This combination of fuel and hot, compressed air causes combustion. On the power stroke, combustion moves the piston down; then on the exhaust stroke, the piston moves up. Before it closes the intake ports, however, blown air forces exhaust gases out the open exhaust valve. Finally, the piston removes any remaining exhaust gases through the exhaust valve.

Two strokes complete the firing cycle as follows:

- 1st stroke, up—exhaust, air intake, and compression; and
- 2nd stroke, down—fuel injection, combustion, and power.

A two-stroke diesel fires on every revolution of the crankshaft. A four-stroke diesel fires on every other revolution of the crankshaft.

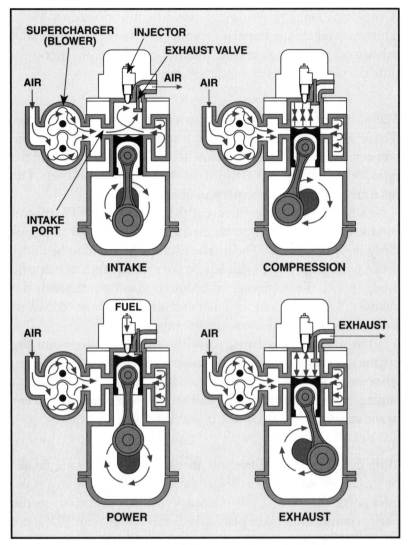

Figure 10. Schematic drawing of a two-stroke/cycle diesel engine

In a two-stroke engine, a power stroke occurs each time the piston reaches TDC. It therefore theoretically produces almost twice the power of a four-stroke engine. Further, a two-stroke engine quickly responds to throttle changes. It also weighs less per horsepower than a four-stroke engine. In spite of the differences between a two-stroke and a four-stroke engine, however, engine choice depends largely on the buyer's preference.

Two-Stroke Power

Concurrent Events in Two-Stroke Engines

A two-stroke engine fires every time the piston reaches TDC. Thus, several things happen at once. For example, intake and exhaust occur at the same time, and the power stroke occurs every time the piston moves down the cylinder.

Power, Exhaust, and Intake Stroke

Manufacturers must use forced-air induction on a two-stroke diesel. To see why, first think of the piston going down the cylinder on the power stroke. Just before the piston uncovers the intake ports in the cylinder wall, the exhaust valves at the top of the cylinder open. The open exhaust valves release pressure inside the cylinder.

At BDC, the piston uncovers the intake ports. The piston cannot, however, draw in air because it is at BDC and has momentarily stopped moving. The intake air, therefore, has to be under pressure to get into the cylinder. So once the piston uncovers the intake ports, the engine-driven blower forces air through the cylinder. This air fills the cylinder and pushes the burned exhaust gases through the still-open exhaust valves.

The amount of exhaust gases the fresh air forces into the engine's exhaust manifold is measured in terms of *scavenging efficiency*. A two-stroke engine should have about 98 percent scavenging efficiency. That is, intake air should move about 98 percent of the exhaust gases out of the cylinder.

Compression and Power Stroke

With the cylinder full of fresh air, the piston moves up the cylinder on the compression stroke. Before the piston completely covers the inlet ports, the exhaust valves close. When the piston covers the ports, compression takes place. With the piston near TDC, the injectors supply fuel, and the power stroke begins as the piston starts down.

Cutaway of a Two-Stroke Diesel

Figure 11 is a cross section of a two-stroke-per-cycle diesel engine. The manufacturer arranged this engine's pistons and cylinders in a V-shape. Several cylinders make up each side of the V.

The drawing shows the fuel-injection system and the ports in the cylinder wall on the left. It also shows the exhaust valves and a cutaway of the piston in the cylinder on the right. In this engine, each fuel injector contains a high-pressure plunger pump operated by a cam and a rocker arm. This type of injector does not require a separate fuel-injection pump.

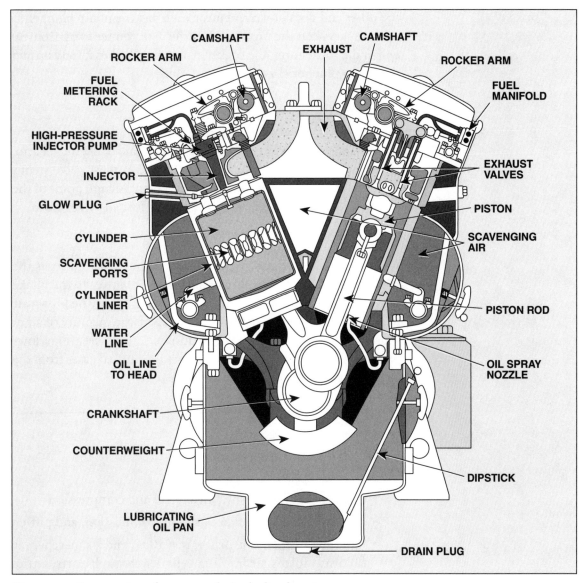

CAMSHAFT   CAMSHAFT

ROCKER ARM   EXHAUST   ROCKER ARM

FUEL
METERING
RACK   FUEL
MANIFOLD

HIGH-PRESSURE
INJECTOR PUMP

INJECTOR   EXHAUST
VALVES

GLOW PLUG   PISTON

CYLINDER   SCAVENGING
AIR

SCAVENGING
PORTS

CYLINDER
LINER

WATER
LINE   PISTON ROD

OIL LINE
TO HEAD   OIL SPRAY
NOZZLE

CRANKSHAFT

COUNTERWEIGHT

DIPSTICK

LUBRICATING
OIL PAN

DRAIN PLUG

Figure 11. Cross section of a two-stroke/cycle diesel engine

Cam, Rocker Arm, and Plunger Pump

A *cam* is a steel disk with an eccentric shape. It is not round, but elliptical (teardrop-shaped), with a high point (fig. 12). The cam is mounted on a *camshaft*, which the engine rotates as it runs.

The camshaft rotates in such a way that the cam's high point lifts one end of a spring-loaded *rocker arm*. Like the rockers on a rocking chair, a rocker arm moves up and down. (You can see a diagram of the cam and rocker arm that operate the injector on the left side of figure 11.) When the cam lifts one end of the rocker arm,

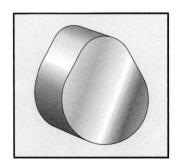

Figure 12. Cam

23

the other end goes down and pushes against the pump plunger to operate it. When the high point of the cam rotates away from its end of the rocker arm, the rocker arm spring pushes its end up and the pump plunger closes.

Cam, Rocker Arm, and Exhaust Valves

As shown on the right side of figure 11, cams and rocker arms also operate the exhaust valves. The cam lifts one end of the rocker arm, causing the other end to press down on the exhaust-valve stems. This action opens the exhaust valves. When the high point of the cam rotates off the rocker arm, the spring-loaded valves close.

Blower and Intake Ports

Referring again to figure 11, note the chamber that encloses the two banks of cylinders. It directs air from the blower to the intake ports around the cylinder walls. The drawing shows the blown air in gray. The blower (not shown) supplies a large quantity of air at a pressure a little higher than atmospheric pressure. This low-pressure air is scavenging air. It removes the burned gases from the cylinders and cools the internal engine parts.

To summarize—

Two-stroke firing cycle

- 1st stroke, up—exhaust, air intake, and compression
- 2nd stroke, down—fuel injection, combustion, and power
- On the upstroke, the piston is at BDC and a supercharger (blower) forces air into the cylinder through ports; at the same time, the blower scavenges exhaust gases through the open exhaust valve.
- On the compression stroke, the piston moves up, the exhaust valve closes, and the piston closes the intake ports. The piston compresses the trapped air and the injector sprays fuel into the cylinder just before the piston reaches TDC.

Diesel Fuel

Low-speed diesel engines operate on almost any liquid fuel, from kerosene to crude oil. Modern, high-speed diesel engines, however, require a lightweight fuel oil. High-speed diesels run so fast that the fuel has a shorter time to burn inside the cylinder. As a result, the weight, or density, of the fuel has to be relatively light.

High-speed engines require diesel fuel with a *specific gravity* of about 0.82 to 0.89 (41° to 27°API). Other diesel engines can use fuel with a specific gravity of about 0.91 (24° API). Specific gravity is the ratio of the weight of one volume of liquid (diesel oil, in this case) to the weight of an equal volume of water. Water has a specific gravity of 1. Thus, a fuel with a specific gravity of less than 1 weighs less than water.

Many years ago, the oil industry established *API gravity* as a density measure for oil and oil products. API gravity is given in degrees API (for *American Petroleum Institute*). The API sets standards, recommends practices, and issues bulletins on all phases of the oil industry.

Two equations show the relation between API gravity and specific gravity—

API gravity = (141.5 ÷ specific gravity) − 131.5   (Eq. 1)

Specific gravity = 141.5 ÷ (API gravity + 131.5)   (Eq. 2)

For example, suppose you had a diesel fuel with a specific gravity of 0.88. What is this fuel's API gravity? The following equation gives the answer:

API gravity = (141.5 ÷ 0.88) − 131.5
= 160.8 − 131.5
API gravity = 29.3 degrees

Specific Gravity and API Gravity

On the other hand, if you had a fuel with an API gravity of 38.2 degrees, its specific gravity would be calculated as follows:

Specific gravity = 141.5 ÷ (38.2 + 131.5)
              = 141.5 ÷ 169.7
Specific gravity = 0.834

Note that the lower a liquid's specific gravity, the higher its API gravity. Thus a lightweight liquid has a low value for specific gravity and a high value for API gravity.

Fuel Quality

The quality of the fuel a diesel engine burns makes a difference in how well it runs. It is therefore important to know the quality of the fuel. Properties that affect a fuel's quality include—

- volatility,
- amount of carbon residue,
- viscosity,
- sulfur content,
- ash, sediment, and water content,
- flash point,
- pour point,
- acid corrosiveness, and
- ignition quality and cetane number.

Volatility

Volatility describes a fuel's ability to change from liquid to vapor, or to vaporize. Testing laboratories measure the volatility of diesel fuel by its 90-percent distillation temperature.

Distillation is the process of driving gases from a liquid by heating it. The 90-percent distillation temperature is the temperature at which 90 percent of the fuel sample has boiled away. The lower the 90-percent distillation temperature, the higher the fuel's volatility.

Small diesels do not run at the high temperatures of large diesels. They therefore require a more volatile fuel. Small diesels running on a fuel with proper volatility have low fuel consumption, low exhaust temperature, and produce a minimum amount of smoke.

Carbon residue is the carbon left after a fuel evaporates. Labora- tory technicians find how much carbon residue is in a fuel by heating it until the volatile part of the fuel completely evaporates. The amount of carbon residue indicates how much carbon the fuel will deposit on an engine's parts. Too much carbon can make an engine run inefficiently: its fuel consumption goes up, the carbon creates hot spots that cause problems, and the engine may not run well. Maximum allowable carbon residue in fuels is 0.1 percent.

Amount of Carbon Residue

Viscosity is a fuel's resistance to flow. A high-viscosity fuel flows more slowly than a low-viscosity fuel. Laboratories measure viscosity by noting the time, in seconds, it takes a given volume of fuel to flow through an *orifice* (an opening) of a given diameter. The shorter the time, the lower the fuel's viscosity.

Viscosity

To measure a fuel's viscosity, technicians often use a *Saybolt viscometer* with an orifice of a universal (standard) size. They measure the number of seconds that it takes a fuel to flow through the universal orifice. Technicians therefore express the viscosity as so many *Saybolt seconds universal* (SSU). The more viscous a fuel is, the longer it takes to flow through the orifice. Thus, fuels with high viscosity have higher SSU values than fuels of low viscosity.

Viscosity influences the fuel's lubricating properties. For example, a thicker (more viscous) fuel tends to lubricate better than a thinner (less viscous) fuel. Also, a more viscous fuel is better at reducing friction and wear on moving parts than a less viscous fuel. What is more, a more viscous fuel is not as likely to leak from threaded connections as a thinner fuel. Keep in mind, too, that fuel is the only source of lubrication for the plungers and barrels in fuel-injection systems, so the fuel's viscosity has to be high enough to provide good lubrication to injection systems.

Sulfur in diesel fuel burns to produce corrosive gases. When an engine operates under a light load, its operating temperature drops. At the lower temperature, corrosive gases in the cylinder turn to liquid. This liquid is very corrosive to engine parts, especially the exhaust system. Thus, corroded exhaust-system parts may show that a problem exists with other parts as well. To reduce such corrosion, fuel specifications limit sulfur content to 0.5 percent.

Sulfur Content

Ash, Sediment, and Water Content

Ash, sediment, and water in a fuel can create problems. Ash and sediment are abrasive and quickly wear out the engine. Sediment may clog the fuel system. Water, especially salt water, in a fuel corrodes engine parts and accelerates wear. Maximum permissible ash content is 0.01 percent. Maximum permissible water-and-sediment content is 0.05 percent.

Flash Point

The *flash point* of a fuel is the temperature to which the fuel must be heated to give off enough flammable vapors to flash (momentarily ignite) when touched by a flame. A fuel with a low flash point, such as gasoline, is dangerous to store and handle. It gives off enough flammable vapors to flash at a low temperature—at or below normal room temperature. Specifications state that the flash point for diesel fuel can be no lower than 150°F (65.6°C). This flash point indicates that the fuel must be heated to 150°F (65.6°C) before it gives off enough vapors to flash.

Pour Point

The *pour point* is the temperature at which a diesel fuel thickens and ceases to flow. At or below a fuel's pour point, you cannot easily pour the fuel out of a container; it is like gelatin. Also, you cannot easily pump a fuel at temperatures below its pour point. The maximum pour point for diesel fuel is 0°F (–17.8°C).

The pour point affects the engine's start-up in cold weather. When fuel reaches a temperature below its pour point, it cannot flow through fuel lines. As a result, the engine cannot start. The engine operator must consider the pour point of a fuel when storing it for transfer to the engine. Temperatures at or below the pour point prevent the fuel from flowing from the storage tank to the engine.

Acid Corrosiveness

The *acid corrosiveness* factor indicates the amount of acids in a fuel that cause corrosion. Refiners must keep the amount of acids low. Otherwise acids can damage metal surfaces both in a storage tank and in the engine. As long as rig operators purchase their diesel fuel from reputable suppliers, they usually do not have to worry about acid levels being too high. Reputable refingers and suppliers strive to keep acids in fuel at very low levels.

Ignition Quality

Ignition quality describes a fuel's ability to ignite when the injector sprays it into the cylinder of hot compressed air. A fuel with good ignition quality ignites readily, with little delay. Conversely, a fuel with poor ignition quality ignites more slowly.

A fuel's ignition quality determines how easily a cold engine starts. Moreover, it determines the kind of combustion that occurs in the engine. A fuel with good ignition quality operates the engine smoothly and quietly.

Cetane Number

One measure of ignition quality is a fuel's *cetane number*. The cetane number is the percentage of cetane in a mixture of cetane and alpha-methyl-naphthalene. Cetane and alpha-methyl-naphthalene are hydrocarbons produced from tar oil. (A *hydrocarbon* is a substance made of hydrogen and carbon.) Cetane has excellent ignition quality, while alpha-methyl-naphthalene has poor ignition quality. A laboratory tests a diesel fuel's ignition quality by comparing it to the ignition quality of a mixture of cetane and alpha-methyl-napthalene.

The ignition quality scale runs from 0 to 100. Pure alpha-methyl-naphthalene corresponds to 0. Pure cetane corresponds to 100. A fuel with a cetane number of 48, for example, has the same ignition quality as a mixture of 48 percent cetane and 52 percent alpha-methyl-naphthalene. High-speed diesel engines require fuel with a cetane number of about 50.

Determining Cetane Number

Technicians determine the cetane number of a fuel by testing it in a special, single-cylinder engine with a variable compression ratio. *Compression ratio* is the volume of air in the cylinder before the piston compresses it compared to the volume after compression.

For example, imagine a cylinder with a volume of 48 cubic inches (in.^{3}) or 786 cubic centimetres (cm^3) before the piston moves up into it. Now imagine that the piston moves up in the cylinder and compresses the cylinder's volume to 3 in.^{3} (49 cm^3). In this case, the engine's compression ratio is 16 to 1 because 48 is to 3 as 16 is to 1 (786 is to 49 as 16 is to 1).

To inject fuel into a cylinder, the engine actuates a *fuel injector*. A short time passes, however, before the fuel actually gets into the cylinder for ignition. This period is the ignition-delay period. In a cetane number test, technicians measure the ignition-delay period at different compression ratios on the test engine. The ignition-delay period decreases as the compression ratio increases.

Testing personnel measure the delay period from the moment the fuel-injection valve leaves its seat until fuel ignition produces a measurable pressure rise in the cylinder. The test uses an ignition-delay period of 13 crank-angle degrees as a reference, which means that the engine crankshaft moves 13 degrees between the time the injector valve unseats and the pressure rise occurs.

Fuels with good ignition qualities require lower compression ratios for the 13-degree ignition delay and have higher cetane numbers. Fuels with poor ignition qualities require higher compression ratios for the 13-degree ignition delay and have lower cetane numbers.

Effects of Unsatisfactory Fuel

A diesel fuel that fails to meet specifications may harm the engine or reduce its efficiency. For example, using a fuel with—

- low volatility reduces the maximum power output, increases fuel consumption, gives a smoky exhaust, and makes cold starting difficult.
- high carbon residue deposits carbon and gummy substances on pistons and cylinder liners. Such deposits may cause the piston rings and valves to stick.
- too high a viscosity may cause a smoky exhaust, excessive wear on injection-pump plungers and barrels, pump leakage, and contamination of crankcase oil by fuel oil.
- too much sulfur, ash, and sediment causes excessive wear on pistons, piston rings, liners, and fuel-injection equipment.
- too high a pour point may make it difficult to start a cold engine.
- too many corrosive acid components causes rapid wear on engine parts.
- poor ignition quality, or a low cetane number, makes it hard to start high-speed engines. In addition, poor ignition quality causes rough, noisy operation.

To summarize—

Diesel fuel

- Properties that affect a fuel's quality include volatility; carbon-residue amount; viscosity; sulfur content; ash, sediment, and water content; flash point; pour point; acid corrosiveness; and ignition quality.
- Volatility should be high for small, cool-running diesels.
- Carbon-residue amount should be no more than 0.1 percent.
- Viscosity should be proper for the engine.
- Sulfur content should be no higher than 0.5 percent.
- Ash, sediment, and water content should be within the maximum limits of 0.01 percent for ash, and 0.05 percent for sediment and water.
- Flash point can be no lower than 150°F (65.6°C).
- Maximum pour point is 0°F (–17.8°C).
- Acid amounts should be low.
- Ignition quality is measured by the cetane number—about 50 for high-speed diesel engines.

Effects of unsatisfactory fuel

- Low volatility reduces power output.
- High carbon residue may cause piston rings and valves to stick.
- Too high a viscosity may cause a smoky exhaust, wear to injection pump plungers and barrels, pump leakage, and contamination of lubricating oil.
- Too much sulfur, ash, and sediment causes wear on pistons, piston rings, liners, and fuel-injection equipment.
- Too high a pour point makes it difficult to start a cold engine.
- The presence of too many corrosive acid components causes rapid wear on engine parts.
- Poor ignition quality (too low a cetane number) makes it hard to start high-speed engines and causes rough, noisy operation.

Fuel Supply Systems

Rig designers can vary the layout of the system that supplies diesel fuel to an engine's fuel-injector valves. All systems, however, share some common features, such as fuel-supply tanks, strainers, filters, supply lines, pumps, and injector valves (fig. 13).

System Using Separate Injection Pump and Day Tank

The fuel supply system in figure 13 is for a diesel engine whose fuel-injector valves require a separate injection pump. Some injectors have a built-in pump and others use a distributor-type pump.

The illustrated system has a main supply tank, a day tank, and several fuel strainers and filters. It also has a transfer pump between the main supply tank and the day tank. (Some supply systems do not need this pump.) The system also has a primary engine fuel pump.

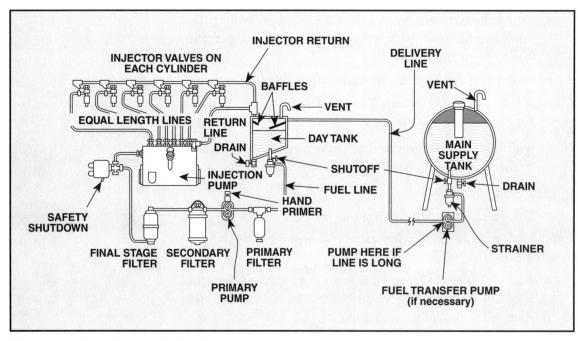

Figure 13. Diesel engine fuel supply system

Main Tanks and Day Tanks

The main supply tank holds a lot of fuel. Either the force of gravity or a transfer pump moves it to the day tank. Typically, a day tank holds enough fuel to run an engine for several hours. The day tank thus supplies the immediate fuel needs of the engine.

On land rigs, which frequently move from site to site, the rig-up crew sets up the engines and small day tanks first. They then fill the day tanks with fuel, start the engines, and use this power to erect the rest of the rig. The crew finally sets the main supply tanks, which provide fuel for the day tanks during regular operations.

Offshore, where it is not necessary to disassemble rig components for moves as it is on land, the rig usually does not need day tanks. Main supply tanks alone provide engine fuel.

On rigs where builders can install the main supply tank above the day tank, they do not need to install a fuel transfer pump. With the main tank above the day tank, gravity moves the fuel.

Pumps Versus Gravity

In the system in figure 13, each fuel-injector valve has its own built-in pump. Built-in pumps eliminate the need for a remote injector pump. The system consists of a supply tank, a primary fuel strainer, a fuel pump, a secondary fuel filter, and fuel lines to the injectors.

System with Built-In Pumps

The gear-type *fuel pump* transfers fuel from the supply tank to the injectors. The injectors do not require that fuel be delivered to them under high pressure. Injector pumps need only moderate pressure to work. The fuel system therefore does not need high-pressure lines or fittings.

Fuel Pump

The pump circulates more fuel through the injectors than they need to spray into the engine. Circulating excess fuel purges (removes) air from the system. Removing air prevents vapor lock when the engine runs on lightweight fuel. Vapor lock occurs when the liquid fuel vaporizes and forms gas bubbles in the fuel system. These bubbles keep the fuel from flowing. The excess fuel also fills, lubricates, and cools the injectors. A line from the injectors returns fuel to the supply tank.

Excess Fuel

Fuel Tank Vents   Referring again to figure 13, note the vents on all the fuel tanks, including the main supply tank. In the fuel system diagrammed in figure 14, the tank has a vent pipe and a vented cap. Vents allow air to flow into the tanks as they empty or fill. When draining a tank, letting in air keeps the tank from collapsing. As the fuel rushes out of the tank, it creates a vacuum (an area of low pressure). If air cannot rush in at the same time to replace the fuel, the higher atmospheric pressure outside the tank could crush it.

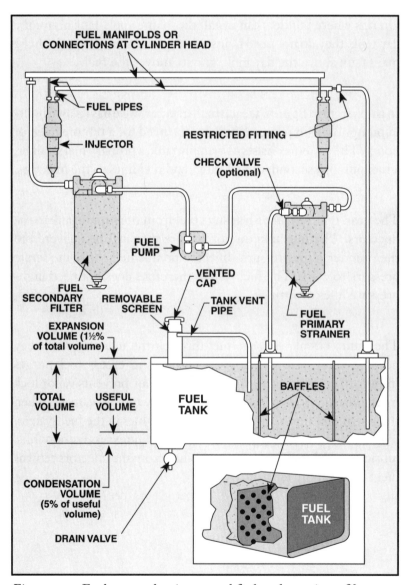

Figure 14.   Fuel system showing vented fuel tank, strainer, filter, pump, and injectors

On the other hand, adding fuel to an unvented tank causes pressure to build up. Pressure could burst the tank. Pressure could also cause a backflow in the fill-up line when a mechanic disconnects it. Fuel would then spew from the line, drenching anyone and anything around it. Therefore, it is important to keep all vents open and all filters and screens clean.

Some installations have automatic vents. These vents stay closed until they sense a vacuum or a pressure in the tank. They then open to relieve the vacuum or the pressure. Automatic vents keep a lot of airborne lint and dirt out of the tank. It is important, though, to keep the automatic opening and closing devices in good working order. Failed vents may interrupt fuel flow to the engine, because no air can get in or out of the tank.

Figure 14 compares the fuel tank's *total volume* with its *useful volume*. Useful volume allows room in the tank for condensate water, which comes from water vapor in the air above the fuel in the tank. The water vapor condenses into liquid water with temperature changes. (Water condensation is like morning dew on a lawn. In the morning, when the temperature drops, the water vapor in the air condenses on the grass as dew.) In a fuel tank, condensate water settles to the bottom. The fuel system withdraws only the fuel above the water.

Useful volume also allows room for the fuel to expand when the temperature rises. In figure 14, the expansion volume is shown to be 1½ percent of the total volume, while the condensation volume is 5 percent of the total.

Total Volume and Useful Volume

Since water is an enemy of the diesel engine, operators always keep the tank that feeds fuel to the engine as full as possible, being careful, though, to allow for expansion. Low fuel levels leave too much room for water vapor on top of the fuel.

Fuel Handling Tip

In figure 13, note the position of the delivery line from the main supply tank to the day tank, exiting the main tank at a point higher than the lowest point in the tank. This arrangement prevents condensate water from being withdrawn with the fuel in the delivery line. The water settles to the lowest point, where personnel can draw it off through the drain. The shutoff valve stops fuel flow and allows a mechanic to work on the strainer and the transfer pump.

Delivery Line Location

Filters, Strainers, and
Centrifuges

Filters and *strainers* remove foreign particles in fuel that cause engine wear. Filters also keep condensate water from entering the engine where it could cause harm. Manufacturers provide stainless steel strainers because brass or copper corrodes in fuel oil.

The strainer on the fill-up line between the main tank and the day tank is usually a 20- to 50-mesh stainless steel screen, which catches any trash that may have gotten into the line. Screen-mesh numbers indicate the number of openings in a linear in. (25.4 mm) of the screen material.

For instance, a square in. (in.^{2}), or a square 25.4 mm (mm^2) of 20-mesh screen has 20 openings along each edge. Thus, there are 400 openings (20 × 20) per in.^{2} (per 25.4 mm^2). Regardless of the number of openings in a screen, the screens should be removed and cleaned in kerosene or other suitable solvent at scheduled times.

Although figure 13 shows a single fuel line for ease of understanding, rig owners usually install two or more strainers and filters in parallel between the main tank and day tank. In figure 13, the strainers and filters are in a series, one after the other in the same line. Installing them in parallel means running two or more lines beside each other (parallel to each other) and installing filters and strainers in each of the parallel lines. Moving fuel through a single line with strainers or filters in series restricts the volume of flow. Thus, a system with only a single strainer or filter line takes too much time to fill the day tank.

Metal-Edge Strainers

Some engine operators prefer *metal-edge strainers* because they are rigid and resist damage. A metal-edge strainer consists of a stack of thick, round, perforated disks separated by thin metal spacers. Each spacer's thickness is 0.0003 in. (0.0076 mm). Assembled on a hexagonal rod, the disks form a cylinder that is placed in a housing. This strainer stops particles thicker than 0.0003 in. (0.0076 mm)—the distance between the disks—against the outside surface of the assembled disks.

Over time, a layer of sludge and dirt accumulates on the outer surface of the strainer disks. This accumulation improves the effectiveness of the filter; however, it also increases the flow resistance. Scheduled maintenance is therefore necessary to remove such accumulations, using the built-in cleaning device. This device is a fixed square rod with cleaning blades. Turning a handle on the strainer's hexagonal rod rotates the disks, and the ends of the stationary blades scrape off the accumulated sludge and dirt. The dirt falls to the bottom of the strainer housing and is removed through a drain.

Tank Filters

Figure 15 shows four tank filters. All have bypass valves that allow fuel to flow around the filter if it becomes clogged. Filters must be cleaned regularly to prevent clogging. If the elements are disposable, they should be replaced regularly, disposing of used filters in accordance with onsite environmental regulations.

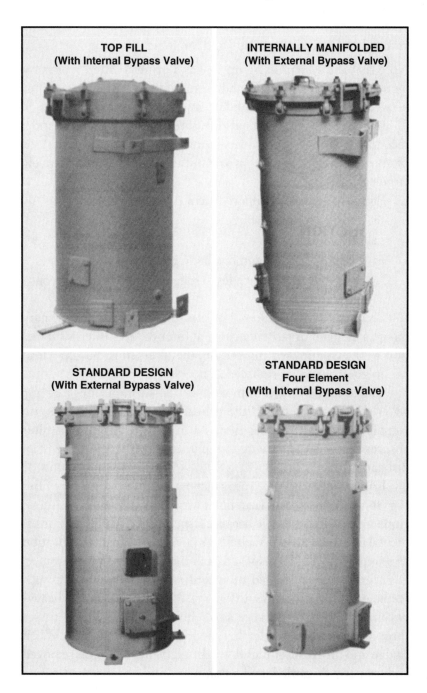

TOP FILL
(With Internal Bypass Valve)

INTERNALLY MANIFOLDED
(With External Bypass Valve)

STANDARD DESIGN
(With External Bypass Valve)

STANDARD DESIGN
Four Element
(With Internal Bypass Valve)

Figure 15. Tank-type fuel filters

37

Bag Filters

A *bag filter* is a woolen small-mesh bag equipped with helical springs that hold it in the shape of a cylinder. The cylindrical surface increases the filtering area, reduces the fuel's velocity, and filters well with only a small drop in pressure. The bag may be cleaned by taking it out and washing it in kerosene or another suitable solvent.

Primary and Secondary Filters and Strainers

Note that the system in figure 13 has a primary filter, a primary pump, a secondary filter, and, sometimes, a final stage filter between the day tank and the fuel-injection pump.

In figure 14, the system also features a primary strainer and a secondary filter. In both installations, the primary strainers or filters protect the engine from harmful abrasives and stop relatively large particles.

Properly designed primary filters
1. allow fuel to flow freely,
2. trap water,
3. remove carbon and insoluble gums, and
4. filter large dirt particles to relieve the load on the secondary filter.

Because a primary filter is on the suction side of the primary pump, a clogged or partially clogged filter reduces or cuts off fuel flow to the engine. For this reason, the filters must be kept clean and in good shape, or replaced as necessary.

The secondary filter removes very small impurities and any water that passes through the primary filter. In some cases, rig owners install still another filter—a final-stage filter. This filter ensures that the fuel flowing to the injectors is virtually free of all dirt and water.

Engine operators often use a disposable filter as a primary filter (fig. 16). It is a perforated can filled with a roll of wool or synthetic cloth. Cotton is unsuitable because it sheds lint. Stringent environmental laws require that such filters be removed from the location after replacing and be disposed of in the correct manner.

Micropore paper folded in accordion pleats often serves as a secondary filtering element (fig. 17). Micropore paper is heavy-duty paper with many very small holes, or pores. Fuel passes through the pores, while the paper catches impurities. When sludge and dirt restrict fuel flow, the element should be removed and properly discarded, and a new one should be installed.

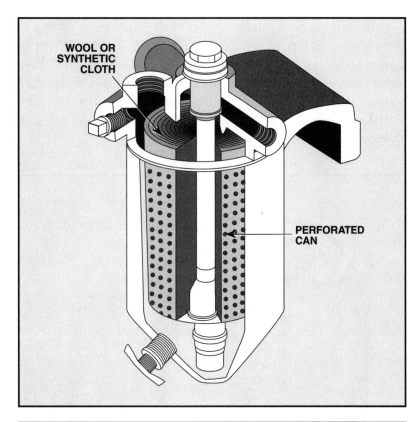

Figure 16. Disposable fuel filter made of wool or synthetic cloth

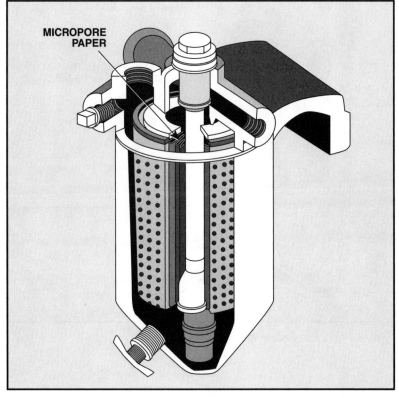

Figure 17. Disposable fuel filter made of micropore paper

Many rig owners employ spin-on fuel filters (fig. 18), which resemble the oil filter on an automobile but are quite different in design. So, do not use an auto oil filter as a spin-on fuel filter. Rigs require specially designed spin-on filters, which must be replaced regularly, using all necessary environmental precautions. A built-in gasket ensures a good seal between the filter and the engine's receptacle.

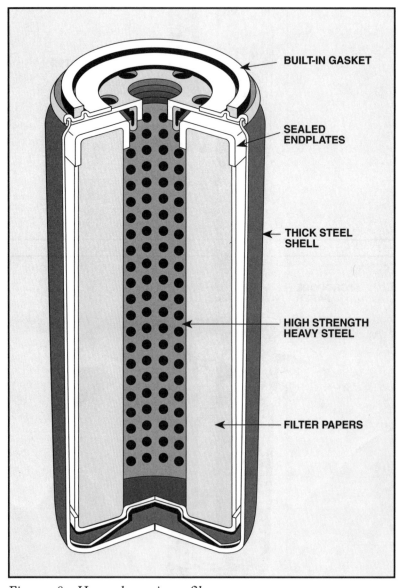

BUILT-IN GASKET

SEALED ENDPLATES

THICK STEEL SHELL

HIGH STRENGTH HEAVY STEEL

FILTER PAPERS

Figure 18. Heavy-duty spin-on filter

Water Separators

An engine operator should not transfer contaminated fuel into the day tank. Dirt and water in the engine's day tank can quickly overload the filters and plug fuel lines.

One method of decontaminating fuel is to use a water separator and a fuel filter. Rig owners usually place the filter between the main tank and the pump that transfers fuel to the day tank. One type of water separator is a *fuel centrifuge*, usually placed between the day tank and the main tank.

The centrifuge spins fuel at a very high speed. As it spins, centrifugal force separates out the particles of dirt and water, which are heavier than fuel. Centrifugal force magnifies the differences in weight between the fuel and the dirt and the water. Centrifugal separation is faster than gravity separation, in which the fuel simply sits until the heavier materials fall out of it.

Adding Water to Fuel for Cleaning Purposes

Particles that weigh about the same as diesel fuel do not readily separate from the fuel, even with a centrifuge. To remove light-weight foreign material, water is often added to the fuel. When water and fuel go through the centrifuge together, the water, which is heavier than the fuel, readily separates out and carries lightweight impurities with it.

Transfer Pumps

A rig's layout determines whether *fuel-transfer pumps* need to be installed. If the builders cannot install the day tanks above the engine's fuel injectors or fuel-injector pumps, the force of gravity cannot deliver fuel to them. In such cases, the rig owner installs fuel-transfer pumps, usually powered by the engine, between the day tank and the engine. To keep from having to shut down the engine to work on a transfer pump, hand- or motor-operated auxiliary transfer pumps are installed.

Pump Sizing

The amount of fuel the main tank and the day tank hold determines the size of the pumps needed to transfer fuel from storage. Most rig owners require that it take no longer than 30 minutes (min) to fill a day tank. If, therefore, an engine has a 450-gallon (gal) or a 1.7-m³ day tank, the transfer pumps must deliver 15 gal per min (gpm) because 450 gal ÷ 30 min = 15 gpm. (This rate is about 0.06 m³/min because 1.7 m³ ÷ 30 min = about 0.06 m³/min.) If the engines have two or more day tanks, the pump may have to be large enough to fill more than one tank at a time.

Pump Location

If the main storage tanks are below engine-room level or below the day tanks, the transfer pumps should be located at the same level as the main storage tanks. Keeping the pumps at the same level as the main tanks ensures that the pumps stay primed (full of fuel).

Locating the pumps at the same height as the main tanks is especially important for centrifugal pumps. Centrifugal pumps cannot lift fuel higher than 20 feet (ft) or 6 metres (m). On the other hand, if the rig uses rotary pumps their location is not as critical. A centrifugal pump has rotating elements like gears or lobes. Two sets of gears or lobes rotate and intermesh within a pump housing. The intermeshing gears or lobes move the fuel.

When a transfer pump is a long way from the day tank, it is important to set the controls of the pump so that it automatically stops after it fills the day tank. This setting cuts down on the danger of overflow. When each day tank has its own pump, the operator can install automatic switches operated by the tank's fuel level.

Fuel Lines

Fuel lines are usually standard steel pipe with screw fittings. Some rig owners use copper or brass tubing with flared fittings for high-grade diesel oils. If the fuel contains sulfur, however, neither copper nor brass fittings should be used, since sulfur corrodes these materials.

Soldered copper fittings should also be avoided, because they are not as strong as flared fittings. Also, they are rigid and tend to break. Fuel flows easier through large-diameter piping than through small-diameter piping. It is important therefore that the diameter of the fuel lines meets the engine manufacturer's specifications. If the fuel lines are too small, they may restrict fuel flow and the engine will not run properly.

Paraffin

Some fuel oils contain *paraffin*. Paraffin is liquid at high temperatures. At low temperatures, however, it is a solid wax. In cool weather, it may drop out of the fuel and stick to the wall of the fuel line. Most engine operators, therefore, use plugged tees, instead of elbows, at all bends in the line. By removing the plug from the tee, personnel can clean paraffin from the lines.

Unions, Gaskets, and Galvanized Pipe

A *union* is a two-part pipe fitting that uses threads and metal gaskets to make a fuel-tight seal (fig. 19). The *gasket* in the coupling makes the leakproof seal. A union makes it easy to connect and disconnect pipe in the middle of the pipe run by screwing or unscrewing the union's threads. Metal gaskets are necessary in this type of union, since the fuel attacks rubber. By the same token, galvanized pipe and fittings should not be used in a diesel-fuel system, because the fuel dissolves the chemicals used in galvanizing and corrosion results.

Figure 19. Stainless steel union connects lengths of pipe

Connections

All fuel line connections coming off the suction side of the pump must be constantly checked to make sure they are tight. Tight connections keep air from entering the line.

Flexible Hoses

Some rig owners use flexible hoses for fuel lines. Flexible hoses do not break or come apart because of engine vibration. Synthetic rubber hoses reinforced with braided steel should always be used, because diesel fuel eats up natural rubber.

Fuel Return Lines

Most diesel fuel systems have return lines from the injectors back to the fuel tanks. These lines recover excess fuel that the system does not inject into the cylinders. Usually, the lines take the excess fuel to the day tanks. This is because the day tanks are usually closer to the engines than the main fuel tanks.

A day tank should never be set more than 6½ ft (2 m) below the engine's primary pump. If the tank is lower, the transfer pump has to develop a lot of suction to lift the fuel. The pump may develop so much suction that a vacuum may be created, causing vapor lock and shutting off fuel flow.

Primary Pump and Injector Pump

Some systems have an engine-driven *primary pump*, which supplies fuel to an *injector pump*. It also has a lever for manual operation. This primary (booster) pump must supply fuel to the injector pump at 10 to 30 psi (70 to 210 kPa) of pressure. Otherwise, the injector pump will not operate properly.

Starting an Engine

To start an engine, the *booster* (priming) *pump* is worked by hand until fuel reaches each injector and bleeds through the injector bleed-off valve. The bleed-off valve opens to let excess fuel drain back to the fuel tank. The bleeding process must continue until all air is expelled from the fuel lines and filters. Air can keep the engine from starting. Even if the engine does start with air in the lines, air knocking can occur.

Air should enter the cylinder only through the intake valve or port. If the injector sprays air and fuel into the cylinder, instead of just fuel, then the cylinder may misfire, or ignite improperly. *Air knocking* is a hammering noise that occurs when the injector sprays a combination of air and fuel into the cylinder. This trapped air can sometimes be removed by cranking the engine with the starter. If this method fails, the booster pump may be manually operated until it pumps out all the air.

Air Knocking

If the engine continues to knock or hammer after such a pumping operation, it must be stopped. The booster pumping operation should be repeated until all air is removed from the fuel. Persistent knocking must be attended to; either there is still air in the system, or there is another problem. It is important to remember in any case that diesel fuel lubricates the fuel system. Therefore, if you operate the system with air in the fuel, it may not lubricate the injector pump enough. A damaged pump can be the result.

Removing Air

To summarize—

Fuel supply systems

- Systems consist of fuel-supply tanks, strainers, filters, supply lines, pumps, and injector valves.
- Main supply tank holds the bulk of the engine's fuel.
- A day tank supplies the immediate fuel needs of the engine.
- Strainers and filters keep the fuel clean.
- Pumps move the fuel through supply lines to the injectors.
- All fuel tanks must be vented to allow air in and out.
- Air and water must be kept out of the fuel lines.
- Fuel dissolves galvanizing chemicals and corrosion results; galvanized pipe and fittings should not be used for fuel.

Fuel-Injection Systems

A diesel engine's fuel-injection system must inject fuel at the right time, inject the right amount, fully atomize it, and inject it in a proper spray pattern.

The injector fully atomizes a fuel when it breaks it into very small droplets that thoroughly mix with the air. The injection system must be efficient and dependable under all speed and load conditions.

A fuel-injection system must—

- accurately meter the fuel,
- inject the fuel at the right time,
- inject the fuel at the correct rate,
- properly atomize the fuel, and
- properly distribute the fuel in the combustion space.

Injection System Requirements

To obtain accurate fuel metering, the injection system must—

- sense the correct amount of fuel to inject for the engine load, and
- inject the same amount of fuel into each cylinder's combustion chamber.

If the system fails in either of these two tasks, the engine will not run correctly.

Accurate Fuel Metering

Injecting fuel at precisely the right moment in the engine's operating cycle is vital. Proper timing gives maximum power from the fuel, good fuel economy, and clean combustion.

Proper Injection Timing

Fuel-Injection Rate   The fuel injection rate is the quantity of fuel the system injects into the combustion chamber during one degree of crankshaft travel. The fuel-injection rate depends on—

- the amount of fuel injected, and
- the length of time (the duration) of the injection.

If the amount of injected fuel decreases, the injection duration must increase. Conversely, if the injection amount increases, the injection duration must decrease. If the injection amount and injection duration do not balance each other, the engine cannot put out consistent power.

Fuel Atomization   Fuel atomization is the breaking up of the fuel stream into a spray of tiny droplets that mix easily with the air in the combustion chamber. Proper atomization ensures that enough oxygen surrounds each small fuel particle for combustion to occur readily. The shape of the chamber determines the type of atomization. Some chambers require a very fine spray, whereas others operate with a fairly coarse spray.

Good Fuel Distribution   Good fuel distribution is obtained when the injectors spray the fuel to all parts of the combustion chamber, where oxygen is available for combustion.

Types of Injection Systems

Engine suppliers provide four kinds of mechanical injection systems for diesels:

1. multipump injectors,
2. unit fuel injectors,
3. distributor injectors, and
4. common-rail or pressure-time injectors.

Diesel-engine fuel systems sometimes use multipump injection systems. A multipump injection system has a plunger pump for each engine cylinder. A plunger pump has a round steel rod, the plunger, which fits inside a steel cylinder, the bushing or barrel. The plunger moves up and down and rotates inside the barrel (fig. 20). Suppliers can mount a pump on each cylinder, or they can combine several pumps into one unit. Each pump in the unit supplies fuel to its own cylinder.

 A plunger pump feeds fuel under high pressure—up to 10,000 psi (70,000 kPa). The highly pressured fuel goes to a fuel-injector valve and spray nozzle on each cylinder. The injector valve opens, and fuel shoots out of the spray nozzle. As the fuel goes into the engine cylinder's combustion chamber, the spray nozzle atomizes it. After fuel injection, the valve closes.

Multipump Injectors

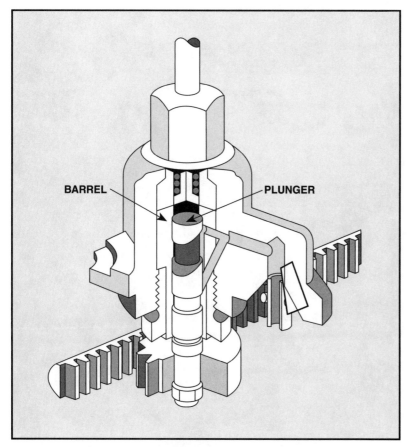

Figure 20. Plunger pump for fuel injection

49

One Pump on Each Cylinder

Big engines sometimes have one pump mounted on each cylinder. Figure 21, for instance, shows one side each of two large V-12 diesels. Each of these engines has a total of twelve pumps.

High-pressure fuel lines carry fuel from each pump to injector-valve and spray-nozzle assemblies on each cylinder. Cams inside the engine operate the pump plungers as well as the engine's intake and exhaust valves. A primary pump (not visible in figure 21) supplies fuel to the plunger pumps.

Figure 21. Multipump injection system with individual pumps placed next to each cylinder

On multicylinder engines, manufacturers often mount several pumps within a single housing, with each pump supplying fuel to the injector valves and spray nozzles on one cylinder. In other words, each cylinder has its own pump but the pumps are put into a single housing, which is mounted at a convenient place on the engine. For example, a unit for an eight-cylinder engine has eight pumps inside one housing, with a single camshaft operating all the pump plungers.

A cutaway of a cam-operated multipump injection system (fig. 22) shows parts of two pumps in a single housing. Note the fuel passage. Fuel, under low pressure, flows through the passage to the individual plunger pumps in the housing. The camshaft of the fuel injector is at bottom. The cam contacts a lifter, which connects to the plunger. A spring fits into the grooves on the plunger. (It is removed in the figure to clearly show the plunger.)

Single-Unit Multiple Pumps

How Cam-Operated Plunger Pumps Work

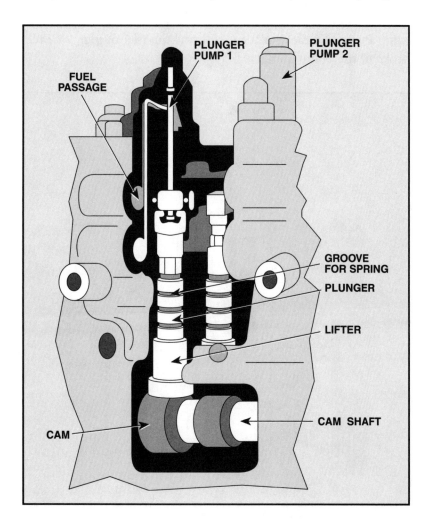

Figure 22. Fuel injector pump

The high point on the cam raises the lifter, which, in turn, raises the plunger. When the cam's high point rotates off the lifter, the spring causes the lifter to go down, which lowers the plunger. The movement of the plunger pumps fuel into the injector.

A cutaway that isolates a single fuel pump (fig. 23) shows the plunger, the barrel, and the rack, which is part of a rack-and-pinion gear. A rack-and-pinion gear consists of a geared, or toothed, bar—the rack—whose teeth mesh with a geared, or toothed, disk—the pinion. Back-and-forth movement of the rack rotates the pinion. Mechanical linkage connects the rack to a governor, which may be installed some distance away.

A governor—

- limits the maximum speed of the engine,
- increases the amount of fuel injected when the driller loads the engine, and
- cuts off fuel to stop the engine.

The engine governor controls the position of the injector pump rack. The rack, in turn, controls the fuel output for each stroke of the pump plunger.

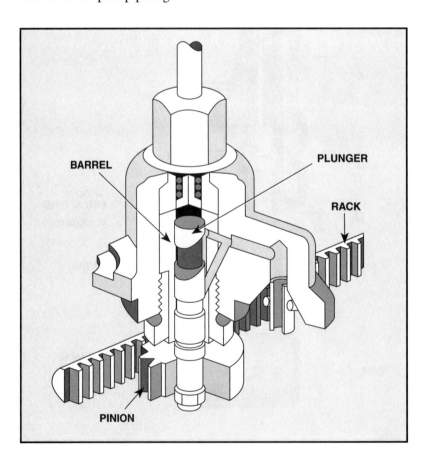

Figure 23. Single injector unit

Figure 24 shows the pinion at the bottom of the plunger. The movement of the rack rotates this pinion, which controls the rotation of the plunger. The plunger's rotary motion measures, or meters, the amount of fuel that the plunger pumps with each stroke.

Counterclockwise rotation increases the amount of fuel pumped. Clockwise rotation decreases the amount of fuel pumped (note that on some injector pumps of this type counterclockwise rotation decreases the amount of fuel pumped and clockwise rotation increases the amount of fuel pumped. It depends on the pump's design). The more load placed on the engine, the more fuel it requires. Conversely, the less load placed on the engine, the less fuel it requires.

Also in figure 24, note the vertical passage with a half-moon-shaped cross section at the top of the plunger. This slot, or passage, joins a recessed area created by reducing the circumference of the plunger. The position to which the rack rotates this part of the plunger determines the amount of fuel sent to the injector valves.

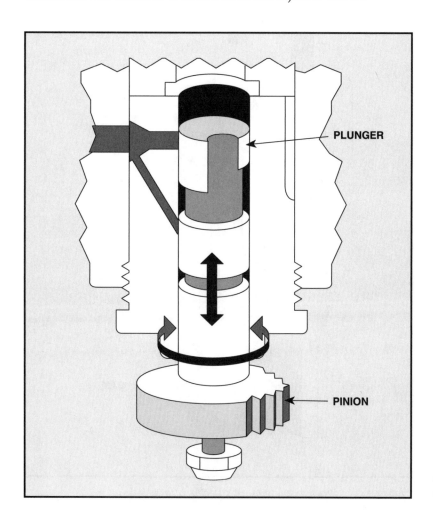

Figure 24. Plunger operation

The cam shown in figure 22 moves the plunger up and down. This up-and-down motion forces the metered amount of fuel into a line to each cylinder.

Figure 25 shows the cam on the camshaft (at bottom) that moves the spring-loaded lifter and connected plunger up and down as the camshaft turns. The cam produces a constant plunger stroke length. Thus, the length of the stroke does not vary the amount of fuel delivered to each cylinder.

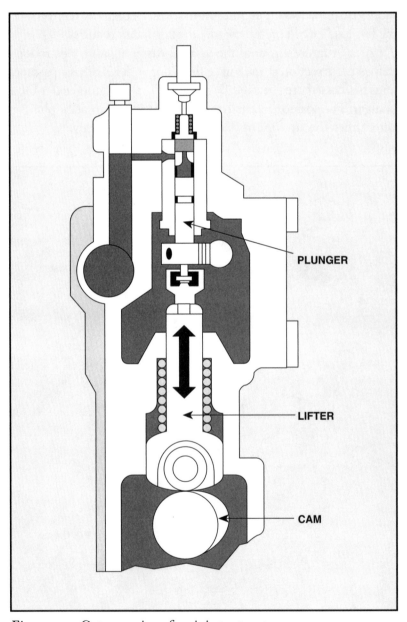

Figure 25. Cutaway view of an injector pump

Note the spring on the lifter below the plunger. This spring forces the lifter and plunger down when the high point of the cam rotates away from the lifter. At top left, you can see the fuel passage into the pump, and fuel, flowing to the top of the plunger through an inlet port.

Figure 26 shows the operation of the pump plunger in detail. Figure 26a shows the plunger at its lowest position in the barrel. Fuel enters the port at left and fills the chamber above the plunger. Fuel also flows down the vertical passage in the plunger (the half-moon-shaped notch) and fills the recess.

In figure 26b, the plunger has moved up, covering the inlet port. It traps fuel in the chamber above the plunger and forces it out the top opening and into the fuel line.

In figure 26c, the plunger is at its highest position. Here, it uncovers the inlet port and allows fuel trapped in the recessed area to return to the fuel inlet. Keep in mind that the inlet fuel is under a lot lower pressure than the fuel on top of the plunger. This lower pressure allows fuel to return to the fuel inlet. When fuel returns to the inlet, no more fuel passes through the outlet at top, and fuel injection stops.

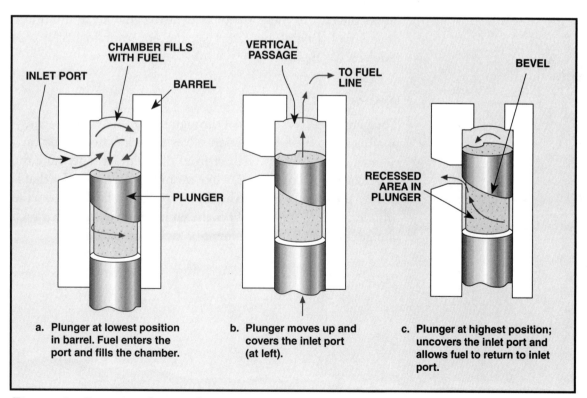

Figure 26. Operation of pump plunger

In figure 26c, note that the top of the recessed area of the plunger is beveled, or cut on a slant. The position of the beveled recess determines how much fuel returns to the fuel inlet. The amount of fuel that returns to the inlet determines how much fuel can go to the engine cylinders.

As an example, when the engine load goes down, the engine needs less fuel to maintain speed. To keep the engine from speeding up, the governor makes the rack rotate the bevel clockwise. Clockwise rotation enlarges the space between the bevel's edge and the port's edge (see fig. 26c). Thus, more fuel returns to the fuel inlet. The plunger, therefore, has less fuel to pump to the injector valve. As a result, the engine's speed does not increase.

On the other hand, when the engine load goes up, the engine needs more fuel to maintain speed. To keep the engine from slowing down, the governor makes the rack rotate the bevel counterclockwise. Counterclockwise rotation shrinks the space between the bevel's edge and port's edge so that less fuel returns to the fuel inlet. The plunger therefore has more fuel to pump to the injector valve. As a result, the engine's speed does not decrease.

Once the plunger delivers fuel to the injector valve, it returns to its lowest position and the process starts again. The governor moves the rack to rotate the plunger as the engine requires more or less fuel. Engine operators set the governor to perform as required.

Injection Lines

The pump plungers send fuel through heavy steel tubing—injection lines—to the fuel-injection valves and spray nozzles on each cylinder. That is, lines farthest from the cylinder determine the length of all the lines. Thus, a line running from a cylinder that is close to the injector is cut to be as long as the line farthest from the cylinder. Having injection lines the same length ensures that each plunger pumps the same amount of fuel to each cylinder.

Injector Valves

A cross section of an injector valve and spray nozzle in closed position (fig. 27) shows its parts. Injected fuel pressure on the upper end of the valve forces the valve off its seat, allowing the nozzle to spray fuel into the engine's combustion chamber. After injection, the spring forces the valve closed while combustion takes place.

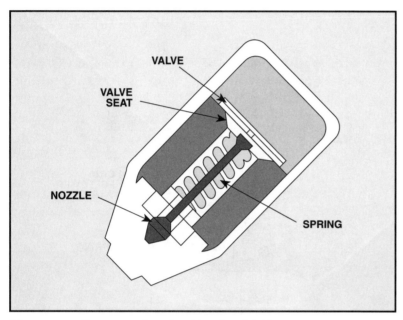

Figure 27. Spray nozzle valve (closed position)

Today, the drilling industry uses mostly unit fuel-injection systems for its engines. Regulatory agencies have passed stringent emission laws, and engineers have added sophisticated electronic controls to engine fuel systems.

    Electronic control of unit injection systems offers—

- precise injection timing and metering, which reduces emissions, and
- instantaneous monitoring of the system, which allows quick response to problems.

Unit Fuel Injection

Unit Fuel-Injector Assembly

A unit fuel-injector assembly (shown in fig. 28 and diagrammed in cutaway fashion in fig. 29) combines an internal high-pressure pump, an injector valve, and a spray nozzle. Fuel enters one of the threaded connections on the left side of the unit and returns to the primary fuel pump from the other connection.

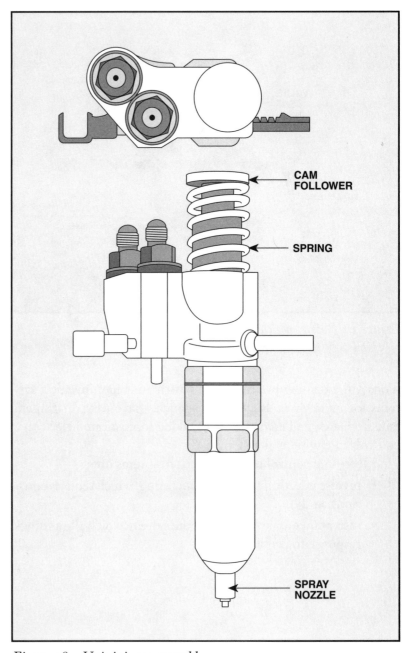

Figure 28. Unit injector assembly

58

A spring surrounds a cam follower, which rests against a cam and follows the shape of the cam as it rotates. In a unit injector, the high point of the cam pushes the cam follower down. The follower, in turn, actuates the pump plunger.

The rack operates the pinion to rotate the pump plunger. Although the teeth that move the pinion are not visible in the side view of the rack, they can be seen in the top view.

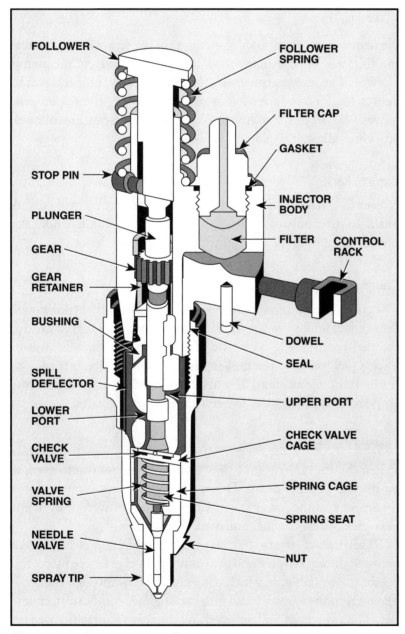

Figure 29. Cutaway view of a unit injector pump

Pump Action

The action of this pump is similar to that of the multipump system. A plunger inside the unit pumps fuel to the built-in spray nozzle. And, as in the multipump system, the pump barrel, or bushing, contains inlet ports that the plunger opens and closes as the rack rotates the plunger. Rotation controls the quantity of fuel pumped on each stroke.

Fuel Delivery

To deliver fuel, a camshaft operates a rocker arm, which contacts the follower on the injector. The follower contacts the pump plunger. The rocker arm pushes the follower and plunger down to deliver fuel. To reduce or shut off fuel delivery, the rocker arm moves away from the follower. A spring then pushes the follower up, which allows the plunger to move up.

Control Rack

A lever linked to the engine governor actuates the control rack. A qualified mechanic can adjust each injector lever independently, thereby obtaining the desired setting for each cylinder.

Continuous Fuel Flow

The engine fuel pump supplies each unit injector with a continuous flow of fuel under low pressure. Although not shown in figure 29, the fuel-return outlet is next to the fuel inlet, which is shown. Excess fuel passes to the fuel-return line, which directs it back to the fuel tank. Continuous flow of fuel through the injector prevents air pockets in the system and cools the injector's parts.

Injector Operation

When the rocker arm (not shown) pushes the follower down, it pushes the plunger down. Downward plunger movement closes the upper and the lower ports. Continued downward movement puts pressure on the fuel under the plunger.

This pressure opens the check valve, and fuel enters the passage between the check valve and the spray tip. Fuel pressure forces the needle valve off its seat, which allows fuel to flow through the small openings in the spray tip and into the engine's combustion chamber. The amount of fuel pumped and the type of spray tip used on the injector nozzle determine the engine's power.

A distributor injection pump is a single pump that sends fuel under high pressure to each injector (fig. 30). It has a plunger into which the manufacturer cuts a channel, or slot. The plunger rotates and the channel distributes the fuel through an outlet to each injector. It distributes the fuel to each injector in the proper firing order.

Distributor Injection

Firing Order

Firing order is the numerical sequence in which combustion occurs in each engine cylinder. For example, in an eight cylinder engine, the manufacturer may set the firing order to be 1–8–4–3–6–5–7–2. Such an order means that combustion occurs in cylinder 1 first, then in 8, then in 4, and so on, until combustion occurs in all eight cylinders. Then the sequence starts over.

Pump Operation

The plunger rotates continuously while moving up and down. Plunger rotation and vertical movement line up ports for fuel metering and distribution. Fuel enters the distributor through inlets not visible in figure 30. Fuel fills the fuel sump, ports, and the cavity between the top of the plunger and the bottom of the delivery-valve assembly.

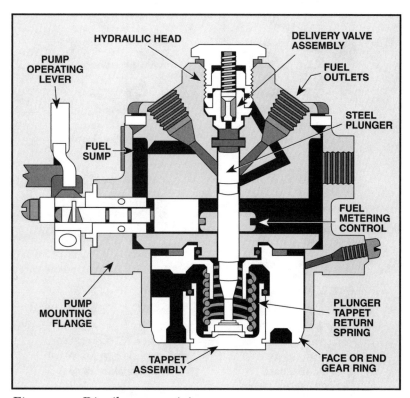

Figure 30. Distributor-type injector pump

Intake

The drawing labeled "intake" in figure 31 shows the fuel on top of the plunger and below the delivery valve. Since the plunger is down, the fuel port is open, while the delivery valve is closed. The upside-down T shape shown in the plunger represents the channel that allows fuel to pass into the plunger.

Beginning of Delivery

At the beginning of delivery, as shown in figure 31, the rotating plunger moves upward, closing the fuel port and putting pressure on the fuel below the delivery valve. This pressure causes the valve to begin opening. At this point, the channel in the plunger has rotated to the left, or clockwise.

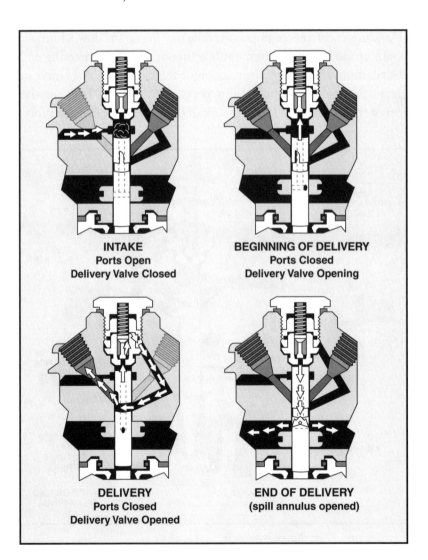

INTAKE
Ports Open
Delivery Valve Closed

BEGINNING OF DELIVERY
Ports Closed
Delivery Valve Opening

DELIVERY
Ports Closed
Delivery Valve Opened

END OF DELIVERY
(spill annulus opened)

Figure 31. Operation of distributor-type injector pump

Delivery

At delivery, the plunger shown in figure 31 has moved up to close the fuel port, and fuel pressure has opened the delivery valve. Arrows indicate the path of the fuel through a passage to the plunger, through the channel in the plunger, and into the fuel outlet. The manufacturer times the channel, or distribution slot, to line up with the outlet to the proper cylinder in the firing sequence.

End of Delivery

At the end of delivery, fuel pressure forces the plunger to maximum height. At this point, a horizontally drilled hole—a metering hole—in the plunger rises above the fuel metering control sleeve. The metering hole relieves pressure on the fuel. The delivery valve closes and the fuel escapes down the vertical hole in the plunger and into the sump surrounding the metering control sleeve. (A sump is simply a space or an area into which fuel drains.)

Fuel Metering

As shown in figure 31, the position of the control sleeve controls the quantity of fuel delivered on each stroke of the plunger. At the end of delivery, the metering sleeve exposes the horizontally drilled metering hole in the plunger. The center hole in the plunger relieves pressure into the sump surrounding the metering sleeve. Fuel delivery stops, despite continued upward movement of the plunger.

Throttling Back and Shutting Down

When the operator throttles the engine all the way back or unloads the engine, the metering sleeve falls to its lowest position and stays there, as at the intake position in figure 31. The plunger continues to move up and down and rotate.

With the metering sleeve at its lowest position, and the plunger moving up, the plunger's metering hole opens before the upper end of the plunger can cover the fuel inlet port. In this position, no pressure can build up even after the plunger covers the fuel inlet port. The vertical hole in the plunger relieves the pressure. With no pressure to open the delivery valve, the injector delivers no fuel, and the engine shuts down.

Throttling Up

Opening the throttle and applying a load moves the metering sleeve to midposition, as at the beginning of delivery in figure 31. In midposition, the metering sleeve uncovers the hole in the plunger later during the plunger stroke. This action lengthens the effective stroke of the plunger, and the injector delivers fuel for normal operation.

Common-Rail Injection

Figure 32 shows a common-rail injection system. Large, low-speed engines, such as marine diesels, often use common-rail injection. A gear pump picks up fuel from a tank and discharges it at a constant pressure and at a constant delivery rate to a header, or common rail. A header is a pipe that distributes fluid—fuel in this case—from other pipes.

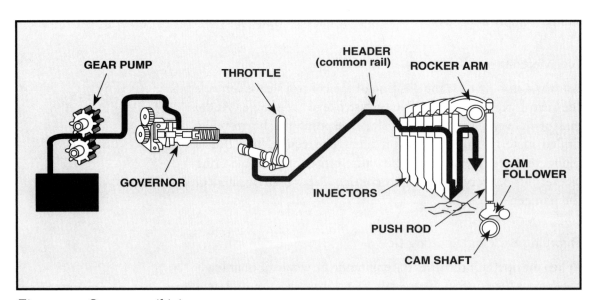

Figure 32. Common-rail injector system

In a common-rail system, each injector is on the common rail. A cam, pushrod, and a rocker arm operate the injectors. A pushrod is a steel shaft that the cam moves when lifting a cam follower. The follower contacts the cam, and when the cam lifts the follower, the follower lifts the pushrod. The pushrod then moves the rocker arm, which operates the injector. A governor and a throttle complete the installation.

Fuel Metering and Injection

Each injector meters and injects the fuel. A fixed-size opening in the injector and fuel pressure carry out the metering process. Fuel metering is based on both time and pressure. The fuel pump and throttle supply the pressure. The interval during which the metering orifice in the injector remains open determines the time for metering. Engine speed establishes the interval and determines the rate of motion of the camshaft-controlled injector plunger. Downward movement of the injector plunger forces the metered fuel into the engine cylinder.

Fuel-Flow Control

A throttle control regulates the fuel flow by increasing or decreasing the amount of fuel that enters the common rail. Increasing the fuel flow increases pressure in the common rail, while restricting fuel flow decreases pressure. The engine operator normally controls fuel flow to the rail by adjusting the throttle control lever. The operator adjusts the throttle until the engine runs a bit too fast. The governor then reduces the fuel flow to reduce engine speed to the desired revolutions per minute (rpm).

The throttle-and-governor system regulates fuel pressure to a point no higher than 150 to 250 psi (1,035 to 1,725 kPa). The amount of pressure and the time at which the plunger opens the injector's inlet port determine the torque and speed of the engine.

To summarize—

Fuel injection systems

- meter the fuel accurately.
- inject fuel at the correct time.
- inject fuel at the correct rate.
- atomize the fuel properly.
- distribute the fuel correctly into the engine's combustion space.

Types of injection systems

- Multipump injectors have a separate high-pressure plunger pump for each individual injector. The pumps may be housed in one common body or individually near the corresponding engine cylinder.
- Unit fuel injectors combine an internal high-pressure pump, an injector valve, and a spray nozzle.
- Distributor injectors have a single pump that sends fuel under high pressure to each injector.
- Common-rail injectors have a pump that picks up fuel from the tank into a header, or common rail; each injector is on the common rail.

Governors

▼
▼
▼

In a diesel engine, the amount of fuel injected into the cylinders controls speed. Many ways exist to regulate the amount of fuel injected. One way (as mentioned earlier) is to use a rack-and-pinion gear, which is usually shortened to just "rack." The rack regulates the position of the plunger in a fuel injector (fig. 33). Pushing the rack to the left increases the quantity of fuel injected. Pulling it to the right decreases the amount. Virtually all drilling rig engines have a governor that moves a rack to regulate speed.

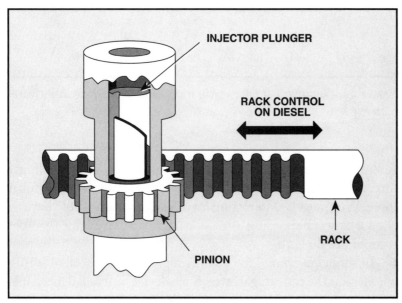

Figure 33. Rack and pinion controls fuel metering.

Centrifugal Force

Governors operate on the principle of centrifugal force. Centrifugal force is created by a spinning object. For instance, if you take a steel bolt, tie a string to it, and twirl it over your head, you can feel centrifugal force trying to pull the string and bolt out of your hand (fig. 34). *Centrifugal force* moves the bolt away from the center of its spin. If the string comes untied, centrifugal force moves the bolt away and into the air.

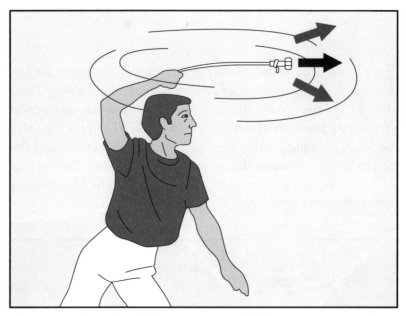

Figure 34.   Centrifugal force moves a steel bolt on a string away from center of spin.

Fly Weights, Springs, and Oil Pressure

Rotating weights (*flyweights*) in a governor produce centrifugal force. A governor must also have a force to counteract centrifugal force. Without a counteracting force, the governor would have to choose between speeding up or slowing down the engine throttle; it could not do both.

In some governors, the compression force of a coiled spring counteracts the centrifugal force. Such devices are called mechanical *spring-loaded centrifugal governors*. In governors that do not use a spring, the force of the engine's oil pressure balances the centrifugal force.

1. *Speed droop* is a decrease in speed (rpm) of the engine from a no-load to a full-load condition. It is usually expressed as a percentage.
2. *Hunting* is a variation, or surge, in engine speed from too fast to too slow. Worn or improperly adjusted governor parts cause hunting.
3. *Sensitivity* is the amount of speed change the engine makes before the governor corrects it.
4. An *isochronous governor* maintains the engine's rpm regardless of the engine's load.
5. Governor power is the force supplied by the governor to overcome resistance created by the fuel system. A governor often needs a lot of force to overcome the resistance put out by mechanical linkages.
6. Promptness is the time the governor takes to react to a change in engine speed.
7. Stability is the ability of the governor to maintain the speed of an engine without hunting.

Terms Used With Governors

Three types of governor are (1) mechanical, (2) hydraulic, and (3) electrically actuated. A fourth type of governor is an overspeed governor, which only keeps an engine from running too fast.

Manufacturers further classify governors according to their functions.

1. Load-limiting governors limit the load an engine can take.
2. Variable-speed governors maintain engine speed at a preset value, from idling to maximum, regardless of load changes. Crane engines often have variable-speed governors, because it is safest for cranes to lift loads at a constant speed regardless of the load's weight.
3. Speed-limiting governors limit only the minimum and the maximum speeds of an engine. The operator manually controls the speeds in between.
4. Constant-speed governors hold the engine at an even speed regardless of the load. Engines that drive generators and therefore require a constant speed often have constant-speed governors.

Types of Governor

Mechanically Actuated Governors

Figure 35 is a diagram of a *mechanical governor*. It shows both the low-speed and the high-speed positions of the flyweights. (Flyweights are simply small, rotating weights.) The engine, through several gears, rotates the bar-shaped yoke the flyweights are mounted on. The inside ends of the flyweight touch the *control sleeve*, which operates the fuel-regulating mechanism.

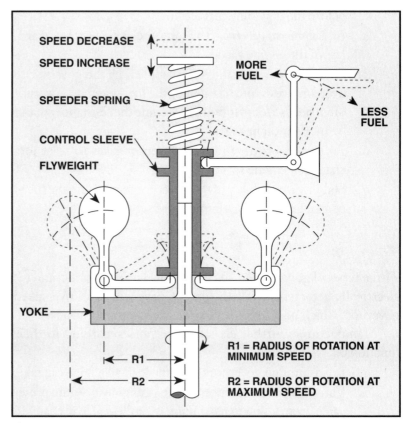

Figure 35. Principle of a mechanical governor (Dotted lines indicate various positions of parts governor will take in course in operation.)

Speeder Spring and Control Sleeve

The *speeder spring* bears against the upper end of the control sleeve. This spring tries to move the sleeve down to increase the fuel supply. The centrifugal force of the rotating flyweights tries to move the control sleeve up to decrease the fuel supply.

Spring Force and Centrifugal Force

In a balanced state, the control sleeve positions the fuel-regulating mechanism to keep the engine speed constant as long as the load does not change. If the load on the engine decreases, engine speed begins to increase. The fuel-regulating setting supplies more fuel than is needed for the load. As engine speed increases, the spinning speed of the governor's flyweights increases. Increased flyweight spin increases the centrifugal force the flyweights produce. The increase in centrifugal force moves the control sleeve up, and this decreases the fuel supply. If the load on the engine increases, the engine slows down. When the engine slows down, the centrifugal force of the flyweights decreases, allowing the speeder spring to move the control sleeve down. Downward sleeve movement increases the fuel supply.

Adjusting a Governor

A variable-speed governor can maintain any speed within the operating range of the engine. With spring-loaded centrifugal governors, the engine operator can change the length of the speeder spring.

Changing speeder spring length changes the governed speed of the engine. A shortened spring increases the centrifugal force required to compress the spring. The engine, therefore, has to run faster for the flyweights to develop enough centrifugal force to overcome the greater spring force. Conversely, a long spring decreases the centrifugal force required to compress the spring, causing engine speed to decrease.

Two-Speed Governors

Two-speed direct-action governors use two springs in the same assembly (fig. 36). Such governors use a soft spring for idling the engine when centrifugal force is small and a stiff spring for higher speeds when the engine is under load. The springs can act either separately or together.

The advantages of mechanical governors are—

- simplicity,
- small size and weight, and
- low cost.

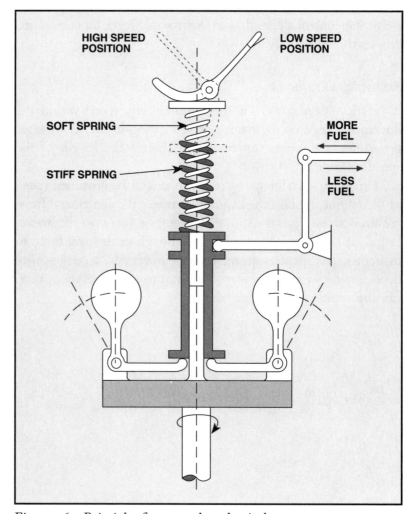

Figure 36.   Principle of two-speed mechanical governor

The disadvantages of mechanical governors are—

- inability to maintain constant speed,
- limited power, and
- poor sensitivity, since the spring not only furnishes the force needed to operate the engine speed control, it also senses speed changes.

In many cases, mechanical governors work well. An exception is where the engines run electric generators. To keep generator output constant, the engines must run at a constant speed. Mechanical governors are not sensitive enough for engines that run generators.

Hydraulically Actuated Governors

Figure 37 diagrams the operation of a *hydraulic governor*. A hydraulic governor is similar to a mechanical governor. The main difference is in the way a hydraulic governor regulates the spring-loaded flyweights. In a hydraulic governor, the control sleeve is not mechanically connected to the fuel-control mechanism. Instead, the control sleeve connects to a pilot valve. Oil under pressure from a pump flows to the pilot valve.

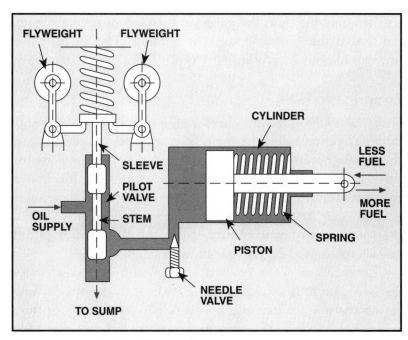

Figure 37. Principle of hydraulic governor

Variations in Speed

When the engine's speed drops below the set speed, the flyweights move inward, lowering the sleeve and pilot-valve stem. Notice the shape of the pilot valve. Its top and bottom are larger in diameter than its middle part. When the pilot valve goes down, the large bottom part of the valve moves away from an inlet to the power piston. With the inlet open, oil flows into the cylinder behind the power piston. The oil forces the power piston to move to the right. This movement increases fuel going to the engine and speeds it up.

When the engine accelerates above the set speed, the flyweights move outward. This movement raises the control sleeve and the pilot-valve stem. The large bottom part of the valve moves to close the inlet to the power piston. Decreasing the amount of oil flowing to the power piston decreases the pressure behind the piston. The piston moves to the left and less fuel goes to the engine, which reduces its speed.

Needle Valve

The *needle valve* in the line to the power piston and cylinder is a throttling device that keeps oil from surging against the piston. A surge causes the engine to overspeed. The governor responds to this by slowing down the engine, often overdoing it. With the engine going too slow, the governor then speeds up the engine. This rapid fluctuation in speed is hunting. The needle valve prevents hunting by preventing surges.

Compensator

Some hydraulic governors have a *compensator*, which prevents hunting by anticipating the engine's return to its set speed. When the engine goes faster than the set (control) speed, the compensator drops the engine's rpm. Normally, operators set the compensator to keep the drop small. With a small speed drop, the compensator governor quickly makes the engine go back to control speed. When the speed drops below control speed, the large power piston quickly returns the engine to control speed (fig. 38).

Compensator governors perform well with any load changes on the engine, from small gradual ones to large sudden ones, as long as the changes are infrequent. Compensator governors do not, however, work well on engines powering devices that constantly change the load on the engine. One example is an engine driving an alternating current (AC) generator. A compensator governor would need constant resetting to maintain a steady engine speed and would therefore be impractical to use in such a situation.

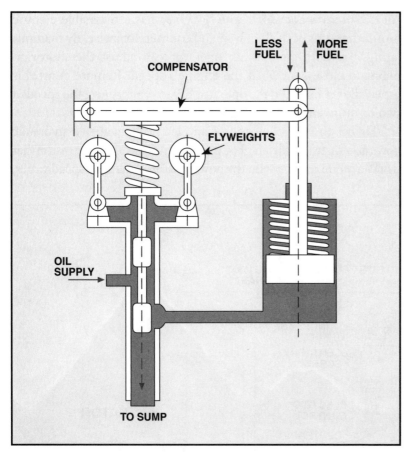

Figure 38. Hydraulic governor with compensating device

Maintaining Hydraulic Governors

Hydraulic governors give years of service if maintained properly. The oil level in the engine oil supply should be checked daily and kept at the proper level; small volumes of oil get a lot hotter than large volumes in a closed system. Also, the oil cannot provide adequate lubrication if it is too low.

Sludge and dirt in a hydraulic governor cause it to operate sluggishly. Therefore, its operating oil should be kept as clean as possible. Usually, a governor's oil is not filtered, so it is easy for dirt to get into the governor housing.

It is important to use the correct type of oil in the governor. If manuals for the rig or the governor do not provide this information, the rig supervisor should be consulted. The oil must contain the right additives and have the correct viscosity. Oil that is too thick causes poor response, while oil that is too thin causes overresponse and too much wear.

Electrically Actuated Governors

An *electrically actuated hydraulic governor* has a reversible electric motor that runs both clockwise and counterclockwise. By manipulating a remote control, an operator can adjust the motor to maintain close control of the engine's speed. Remote control is especially useful when the operator has to synchronize the speed of two or more engines driving generators.

Figure 39 is a schematic of an electrically actuated hydraulic governor. Instead of flyweights, this governor has a loading piston that works in combination with the power piston. As engine speed varies,

Figure 39. Electrically actuated hydraulic governor

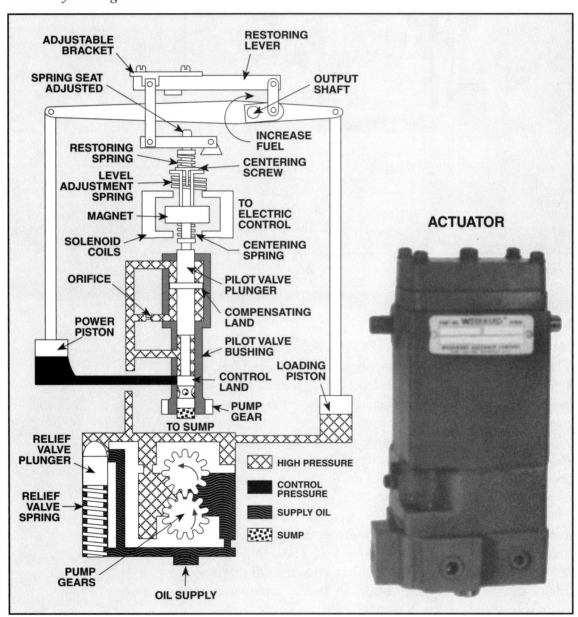

the pilot valve meters the oil behind the power and loading pistons to increase or decrease the engine's speed as required. The motor in the actuator is controlled by the driller on the rig floor.

The driller's remote control panel has a two-way switch. When the driller turns it on, the governor's motor adjusts the engine's speed. When the engine reaches the desired speed, the driller turns the switch off. The driller can tell the engine's speed by looking at a tachometer or frequency meter on the control panel.

By holding the switch in the "lower" position, the driller can slow the engine and even shut it down by leaving the switch engaged. In the "rise" position, the switch increases the engine's speed until it reaches the maximum speed setting on the governor. This maximum speed setting keeps the engine from going faster even if the driller keeps the switch in the "rise" position.

Direct-Acting and Reverse-Acting Electric Actuators

Two types of electric actuators are direct acting and reverse acting. A *direct-acting actuator* increases the engine's speed by increasing the positive voltage going to the actuator and governor; that is, as the engine needs more fuel to go faster, the direct-acting actuator increases positive voltage to make the governor increase the fuel.

A *reverse-acting actuator* increases the engine's speed by decreasing the positive voltage going to the actuator and governor. As the engine needs more fuel to go faster, this actuator decreases the positive voltage to make the governor increase the fuel. An advantage of a reverse-acting actuator over a direct-acting actuator is that if electrical power fails, this actuator causes the governor to deliver a maximum amount of fuel. The governor still, however, holds the engine to its maximum set speed so that overspeeding does not occur. When power fails in a direct-acting actuator, the engine shuts down.

Fuel Modulators

Diesel engines must meet federal and local standards on pollution. One concern is that a diesel may produce too much smoke. If the governor delivers more fuel than air to the engine, it is too 'rich' and the engine smokes too much. To prevent smoking, operators install *fuel modulators* in combination with a governor. The modulator makes the governor increase the fuel supply only at the same rate as the air increase. Such rate control holds down the black smoke from the engine exhaust during acceleration or sudden loading.

Overspeed Governors

Overspeed governors bring an engine to a full stop by cutting off either the fuel supply or the air supply to the engine. They are sometimes called 'trips' because they shut the engine down and protect it from damage if overspeeding occurs. Rig owners usually install overspeed governors along with regular governors to prevent engine damage if the regular governor fails. In cases where the engine does not have a regular governor, an overspeed governor shuts down the engine when the speed increases beyond a safe limit before the driller can control it. Mechanical overspeed governors operate by the centrifugal force of flyweights. Oil pressure operates hydraulic overspeed governors, and electricity operates electrical overspeed governors.

One type of overspeed governor uses a power spring to operate the shutoff control. The installer sets the spring with a latch. If the engine goes too fast, a spring-loaded centrifugal flyweight moves out and trips the latch. When the latch trips, the power spring operates the shutoff control.

To summarize—

Governors

- Almost all engines have a governor that moves a rack to regulate speed.
- Speed droop is the decrease in rpm of an engine from a no-load to a full-load condition.
- Hunting is a variation in engine speed from too fast to too slow.
- Sensitivity is the speed change the engine makes before the governor corrects it.
- An isochronous governor maintains the engine's rpm regardless of the engine's load.
- Governor power is the force supplied by the governor to overcome mechanical resistance in the fuel system.
- Promptness is the time it takes a governor to react to a change in engine speed.
- Stability is a governor's ability to maintain engine speed without hunting.

Types of governors
- Mechanically actuated
- Hydraulically actuated
- Electrically actuated

Mechanically actuated governor
- Flyweights mounted on a gear-driven, rotating yoke are brought to bear against a control sleeve which operates the fuel-regulating mechanism. The speeder spring tries to move the sleeve down to increase fuel, while the centrifugal force of the rotating flyweights tries to move the control sleeve up to decrease fuel.

Hydraulically actuated governor
- A control sleeve connects to a pilot valve; when speed drops, the flyweights move inward to lower the sleeve and pilot valve. When the pilot valve goes down, an inlet opens to let hydraulic oil flow into a cylinder behind a power piston. The power piston moves to increase fuel.
- When speed goes up, the flyweights move outward to raise the sleeve and pilot valve. When the pilot goes up, the inlet decreases the amount of oil flowing behind the power piston. The piston moves to decrease fuel.

Electrically actuated governor
- The engine operator adjusts a switch on a motor to control the engine's speed. The governor has a loading piston in combination with a power piston. As engine speed goes up or down, a pilot valve meters the oil behind the pistons to increase or decrease the speed.

Overspeed governors
- Overspeed governors cut off fuel or air to an engine to stop it; they are available in mechanical, hydraulic, and electrical versions.

Lubrication Systems

A 1,000-horsepower (hp) or 700-kW engine may weigh 20,000 pounds (lb) or 10,000 kilograms (kg) or more. It has hundreds of moving parts that should give thousands of hours of service. In spite of its size and number of parts, a large engine operates on a relatively small amount of oil. A little oil goes a long way, because it has to form only a very thin film between the moving parts to do its job. The thin film of oil reduces the destructive friction that results when parts move against each other.

The *crankcase* is the frame of the engine, and lubricating oil is stored in the bottom of it. In a large engine, the crankcase may contain 100 gallons (gal) or 380 litres (L) of oil; however, only about 5 gal (20 L) of oil forms the film in the engine. When the engine is running, therefore, about 80 percent of the oil remains in the crankcase while the rest flows through the oil filter, the oil cooler, the oil pump, and the lubricating lines.

What Lubricating Oil Does

Lubricating oil in a heavy-duty engine—

- provides a film between moving parts. This film prevents metal-to-metal contact and reduces friction and wear.

- cools the internal engine parts that it touches, such as the underside of a piston or the moving parts of a bearing.

- forms a pressure seal (a barrier) between the combustion chamber above the pistons and the crankcase.

- removes gummy compounds that combustion and heat produce inside an engine.

- operates the governor that regulates the speed of the engine.

- operates safety shutoff controls on industrial and marine engines.

Oil Pumps

As the engine runs, it powers an *oil pump*. The pump is an integral part of the engine. The oil-pumping system forces cool, filtered oil through the engine. It pumps the oil under pressure and at a fast rate of flow. Depending on the engine's size, the oil-pumping system (fig. 40) can circulate all the oil through the engine in a couple of minutes or less.

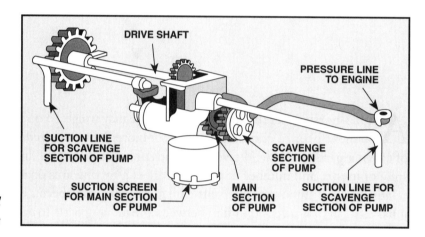

Figure 40. Lubricating oil pump for a diesel engine

Oil Flow Rates

As an engine runs, wear occurs on its moving parts. As wear occurs, the amount of space—the *clearance*—between the moving parts increases. With increased clearance, there is less restriction to the flow of oil between the parts. The oil-flow rate therefore increases, and the pump must be able to maintain this flow-rate increase. If it cannot, the oil pressure goes down and some parts may not get enough oil.

Relief Valves

Most engines have a *relief valve* in the oil-pumping system that maintains oil pressure as the engine parts wear. The pump moves most of the oil to the engine parts. Part of the oil, however, returns to the crankcase through the relief valve. If the engine is running at high speed, a lot of oil goes to the crankcase. At intermediate and low speeds, less oil goes to the crankcase. The relief valve keeps a constant pressure on the oil system whether the engine is running at a low, an intermediate, or a fast speed.

As wear increases, less oil goes through the relief valve, regardless of the engine's speed. When an engine or an oil pump becomes badly worn, most of the oil goes to the engine. A small amount returns to the crankcase through the relief valve; even so, the oil cannot do its job. With large clearances, the pump cannot maintain enough pressure to form a good film. When an engine's oil-pressure gauge shows low pressure at normal operating speed, the engine or the oil pump needs major repair.

As oil circulates through an engine, it picks up dirt, small metal particles, and other foreign material. All can damage the engine. Manufacturers put strainers and filters on most engines to clean the oil. Strainers take out large particles, whereas filters remove small ones.

Oil Strainers and Filters

Strainers

Manufacturers place the oil pump's inlet very close to the top of the oil in the crankcase. They also place on the inlet a strainer that consists of a fine-mesh bronze screen. Some strainers float on the oil's surface. The inlet and strainer are close to the top because most foreign matter is heavier than oil and sinks to the bottom of the crankcase. With the matter on bottom, the pump cannot pick it up and it therefore does not circulate through the engine. Any material that does not sink to the bottom, and is relatively large, is caught by the oil strainer. The oil filter traps the finer particles.

Filters

Manufacturers make many sizes and shapes of oil filter, utilizing such varied materials as paper, wool, cotton, metal, and charcoal. Most often, however, paper and cotton are the preferred filtering elements.

It is important to replace the old oil filter with a new one of the proper size and type. Many filters look the same on the outside but are different on the inside. For example, a fuel filter and an oil filter can be the same size and look the same, but not be interchangeable. Installing a fuel filter in place of an oil filter can cause serious problems. The small holes in the fuel filter do not allow thick oil to flow easily. As a result, much of the oil flows around the filter, and no filtering occurs. On the other hand, fuel passes easily through an oil filter, but the filter's size does not stop pieces of dirt that could harm the fuel-injection system.

Importance of Filtering

Filtering is an important part of the lubrication system. Any hard particle larger than the clearance between two moving parts can wear or damage an engine. An insert bearing resting in a rod cap, for example, is very thin—0.005 in. to 0.015 in. (0.127 mm to 0.381 mm). If the oil carries a particle thicker than the thickness of the oil film coating the bearing, the particle scores the bearing's surface (fig. 41). An oil filter has to catch all such particles.

Some particles are, however, so small that they can circulate in the engine without causing wear. If the oil filter trapped all these tiny particles, it would quickly stop up. The engine manufacturer therefore recommends using a filter that balances engine protection against filter life. Many engines use filters that last as long as 1,000 hours, depending on the output of the oil pump, the number of filter elements, and other factors.

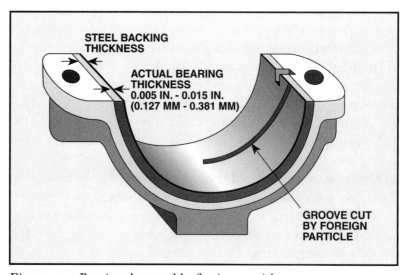

Figure 41. Bearing damaged by foreign particle

Number of Filters

Manufacturers provide heavy-duty engines with two or more filters (fig. 42). A lot of oil passes through such filters. If, for example, an engine can pump oil at 22 gal (0.08 m3) per minute, then it pumps 1,320 gal (5 m3) every hour. If two filters are on the engine, the filters clean 660 gal (2.5 m3) of oil per hour. So, if the engine mechanic changes the filter elements after 125 hours, each will have filtered 82,500 gal (over 300 m3) of oil.

Figure 42.   Oil filter units on a heavy-duty diesel engine

Filter Pressure Gauges

Engine operators use many methods to check whether a filter is still doing its job. One of the best is to have two *pressure gauges*, one on the inlet side (where oil enters the filter), and the other on the outlet side (where oil leaves the filter). The two gauges indicate the resistance that the filter puts on the oil flow.

The gauge on the inlet side should indicate slightly higher pressure than the one on the outlet side. The pressure difference should, however, be no more than 7 to 10 psi (50 to 70 kPa). A 15-psi (100 kPa) difference between the two gauges means that the filter is clogging. When a filter clogs, a motorhand or mechanic should replace it immediately.

Filter Bypass Valves

Oil filter manufacturers add a *bypass* (relief) *valve* to each filter as a safety feature. This valve opens to let oil bypass the filter if dirt blocks it up. Dirty oil provides better lubrication than no oil at all. The bypass valve starts working when the filter puts up more than about 20 psi (140 kPa) of resistance.

Oil Coolers

The oil pump on most engines sends the oil through an *oil cooler* before it goes to the filters. If oil gets too hot, it breaks down and cannot lubricate. Engine manufacturers often use the engine's coolant to cool the oil as well as the engine.

One type of oil cooler is placed near the oil-pumping system (fig. 43). Oil from the pump enters a line at the bottom of the cooler body. The line coils around inside the cooler body and exits from the side. The coolant enters the top of the cooler body; it also exits from

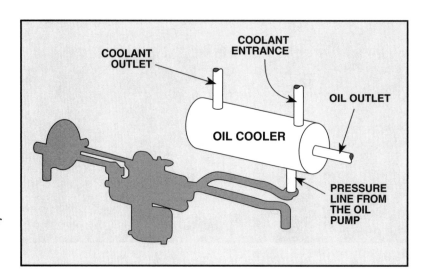

Figure 43. Placement of an oil cooler in lubrication system

the top but from another side. As the hot oil in the oil line contacts the lower-temperature coolant in the cooler body, some of the heat in the oil goes to the coolant, which carries the heat away from the oil.

On some engines, the manufacturer installs a bypass system with the oil cooler (fig. 44). When the engine is cold and the oil is thick (viscous), it cannot flow easily. To warm the oil, the bypass system closes the line to the cooler. At the same time, the system opens a line that bypasses the cooler. With the cooler line closed and the bypass line open, the oil moves through the strainers and filters directly into the engine. When the oil becomes warm enough to flow freely and to need cooling, the system closes the bypass line and opens the cooler line. With the bypass line closed and the cooler line open, the oil flows again through the cooler.

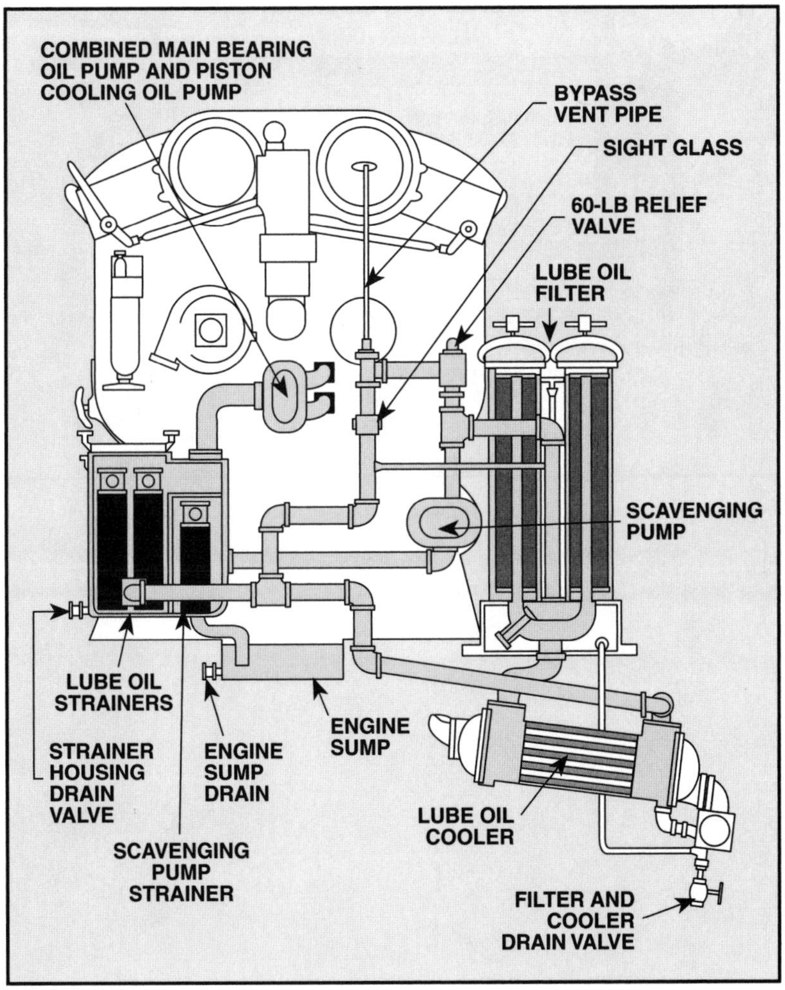

Oil Cooler Bypass System

Figure 44. Lube oil system of a diesel engine showing oil cooler bypass system (striped lines)

87

Engine Oil Supply Areas

Oil goes through a manifold (a pipe with several outlets) to five areas in an engine (fig. 45). These areas include

- the crankshaft, the camshaft, and the pushrod guides;
- the pistons, the piston pins, and the cylinder liner;
- the valve mechanism;
- the timing gear; and
- the fuel-injection system and governor.

If the engine has a turbocharger, the supply system moves oil to it separately. Turbochargers run at very high speeds and therefore need their own oil line. High-speed equipment must have adequate lubrication or it quickly fails.

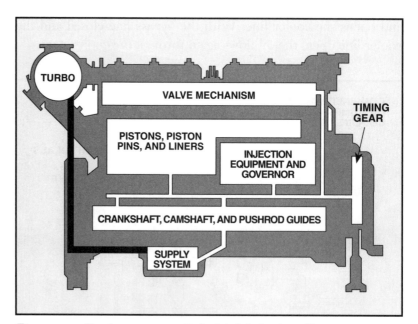

Figure 45. Engine parts supplied with lubricating oil

Crank throws are machined and polished areas on the crankshaft to which the piston rods and piston-rod bearings are attached. Engine crankshaft manufacturers drill holes from each main bearing to the crank throws (fig. 46). Oil flows through these holes (passages) to the piston-rod bearings. Oil also flows to the main bearings through a hole in each bearing.

Crankshaft Lubrication

Part of the oil supplied to each main bearing forces its way along the length of the bearing and then falls back into the crankcase. The other part of the oil goes through the passage in the crankshaft to the adjacent rod bearing. This oil lubricates the piston-rod bearings, the piston wrist pins, and the piston. (*Wrist pins* are hardened steel cylinders that attach the piston to the rod.)

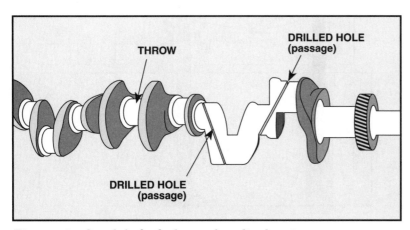

Figure 46.   *Crankshaft of a heavy-duty diesel engine*

A drilled passage in the piston rod carries oil from the piston-rod bearing to the wrist pin. Part of the oil lubricates the wrist pin, but most of it spurts onto the bottom of the piston. This oil cools the pistons.

Piston Lubrication

An engine manufacturer may locate the camshaft above the valves or in the cylinder block. In either case, the camshaft and the cams on it operate the intake and exhaust valves. In most engines, the cam does not directly touch the valve stem. Instead, the cam contacts a *pushrod*, which, in turn, moves a rocker arm (see fig. 9). The rocker arm actually operates the valve. Oil passages supply oil to the camshaft bearings and the pushrod guides. (A *pushrod guide* is a hollow metal cylinder through which the pushrods move.)

Camshaft Lubrication

Rocker Arm and Valve Lubrication

Oil flows through a drilled passage in each rocker arm and lubricates the ends of the rocker arm and the top of the valve stem (fig. 47). Also, a small amount of oil flows along the valve stem to lubricate the valves.

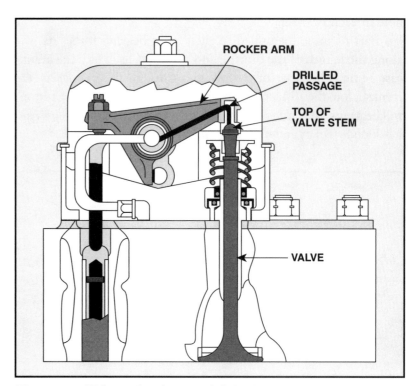

Figure 47. *Valve and rocker arm lubrication*

Timing Gear Lubrication

Diesel engines have a *timing gear train* (a set of gears that drives various engine parts; fig. 48). The crankshaft drives these gears. Engine parts driven by the timing gears include the oil pump, the valve mechanisms, the fuel injectors, and the water pump. Also, in many cases, the timing gears drive hydraulic pumps and other accessories attached to the engine. Oil flows to the timing gears through small lines and passageways, many of which are located in the engine block (fig. 49).

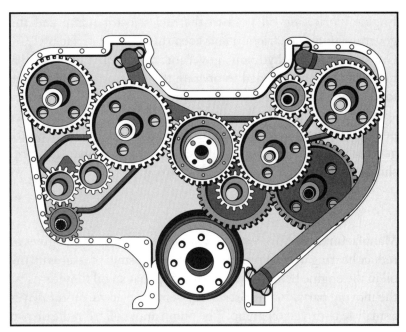

Figure 48. Set of timing gears in a heavy-duty diesel engine

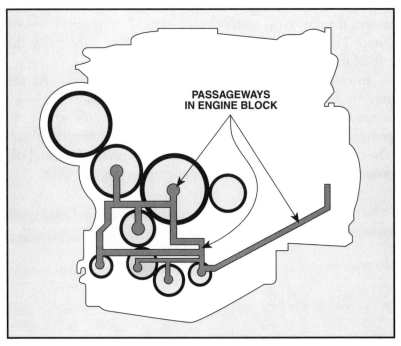

Figure 49. Lubricating oil channels to timing gears

Injector Pump and Governor Lubrication

Engine lubricating oil goes to the fuel injector pump and the governor to reduce friction and keep the rubbing surfaces clean. On engines with a hydraulic governor, the oil pressure provides much of the force needed to operate the governor, reducing the force needed on the governor's shaft. Oil pressure also operates speed limiters and shutoff controls on some engines. A *speed limiter* prevents the engine from running too fast before the oil pressure gets high enough to get oil to all the parts. A safety shutoff control shuts down the engine when oil pressure gets too low.

Prestart Lubrication Systems

Manufacturers install *prestart lubrication systems* on some engines to reduce bearing wear. A prestart lubrication system pressures up the oil in the engine before it starts, ensuring that an oil film forms on the moving parts. In one prestart system, an electric motor drives a small gear-driven oil pump. The pump draws oil from the oil pan and discharges it into the main oil system.

When an operator shuts down an engine, even for a short time, most of the oil drains into the sump, or lowest part, of the engine's oil pan. When the operator activates the starter switch to restart the engine, the prestart system's electric motor begins driving the oil pump. The pump circulates oil into the engine, lubricating the critical moving parts before the engine starts.

In cold-weather operation, an engine operator can rig the pump to circulate oil through heaters. The heaters warm the oil before it goes into the cold engine. Also, the operator can use the pump to fill the engine's filter container after changing the filters. The container may hold as much as 50 gal (200 L) of oil, and the prestart pump makes it easy to handle such a large amount.

Explosion Covers

Fresh air entering through the base of an engine can create a problem when oxygen mixes with the oil in the crankcase. If a hot spot occurs in the engine, the heat can cause the oil and oxygen to explode and damage the crankcase. To prevent such damage, some engines have spring-operated *explosion covers* (doors) fitted to the crankcase. If an explosion occurs in the crankcase, the doors open to release the pressure from the explosion and then slam shut to prevent more air from entering. If the engine operator sees one of the doors fluttering, the operator should notify the mechanic and safely shut down the engine as soon as is practical.

Oil Quality

Modern lubricating oil for diesel engines is better than ever. Improved chemistry keeps oils from oxidizing at high temperatures. Special additives improve the oil's film strength at high bearing pressures. They also keep the oil from foaming, which reduces its ability to lubricate properly. Some additives prevent the oil from failing under conditions of heat and pressure. Others keep the oil thin for cold-weather starting. Modern lubricating oils for diesel engines are high-detergent oils that clean while they lubricate. (A *detergent* is a cleaning material incorporated into the oil.)

The detergent loosens dirt, varnish, sludge, and residues, which then enter the oil. A dispersant in the oil suspends these particles, which makes the oil look dirty. When detergent oil appears especially dirty, this shows that the oil is doing its job by suspending the dirt to keep it from harming the engine.

The lubricating properties of an engine oil practically never wear out. The products of combustion, however, do contaminate the oil as the engine runs and the additives' ability to protect the engine and preserve the oil base wear out. If an engine operator fails to change the oil on a regular schedule, the oil loses its ability to lubricate properly.

Sulfur Content of Fuels and Oil Quality

Impurities in fuel can cause lube oil problems in an engine. One of the worst of these is sulfur. When an engine burns fuel with sulfur in it, an acid is produced which eventually works its way into the lubrication system. Many lube oils have additives that neutralize the acid to keep it from corroding engine parts.

The sulfur content of diesel fuels varies. Thus, the amount of acid produced from sulfur in the fuel also varies. Diesel engine operators should, therefore, adjust the oil-change schedule according to the fuel's sulfur content. The higher the sulfur content, the more often the operator should change the oil.

Using the Proper Oil

When selecting the oil for an engine, the engine's location, its type, and the temperature of its surroundings must be known. The manufacturer's recommendations for the kind of oil to use should be followed.

For example, for warm-weather operations, the manufacturer usually recommends using a heavy-weight (high-viscosity) oil. Conversely, in cold weather, a light-weight (low-viscosity) oil is usually recommended. In cold weather, an engine runs cooler than it does in hot weather. At cool temperatures, a light-weight, low-viscosity oil flows more easily than a high-viscosity oil. Thus, in cold weather, a light-weight oil lubricates the engine better than a heavy-weight oil.

While the viscosity of the oil is important, it is not the only quality to consider. It is essential to use only oil designated for use in diesel engines. Oil manufacturers identify lubricating oils with codes such as API 40-DS-HD. This means that the refiner made the oil to American Petroleum Institute (API) specifications and certifies it to be 40-weight viscosity, to be suitable for diesel service (DS), and to contain high-detergent (HD) additives.

Brands of oil should never be mixed. If it becomes necessary to add oil to the engine to bring the level to normal, the same brand of oil as has been used previously should be used to fill the crankcase. Oil manufacturers use different additives and blend their oils differently, and sometimes the additives are not compatible or the blends may not work correctly if mixed. When switching from one brand of oil to another, the operator should change the filters as well as dispose of all the used oil in an environmentally safe manner.

Detergent Oils

In the old days, an engine mechanic could take a sample of oil from an engine, put it on a finger, and decide whether it needed changing. If the oil looked dark and dirty, it was time to change it.

Today, the high-detergent oils used in diesels get dark almost immediately. The detergent cleans inside the engine and breaks down carbon deposits. As it cleans, the oil carries the dark carbon particles into the filter system. The filter removes particles large enough to harm the moving parts. The small particles, though they are too small to damage the engine bearings, remain in the oil and make it dark. Since an operator can no longer tell by looking at the oil whether it is dirty, it is best to keep accurate records and change it at recommended intervals.

Oil Contamination

Even though an operator changes the oil at recommended intervals, fuel can contaminate the oil and cause it to fail long before a scheduled change. An observant motorhand can easily spot fuel contamination. If the oil level gets higher as the engine runs, and no one has added oil to the engine, then another liquid is getting into the oil. If the liquid is water, the oil has a milky color. If the liquid is fuel oil, the oil pressure gradually drops and the oil looks clear. A mechanic should find the place where fuel is entering the oil system and repair it promptly.

Oil Testing

Operators should take an oil sample from each engine, label it, and send it to a laboratory for testing at least every six months. Among other things, lab tests can show whether personnel are changing the oil frequently enough. They can also show whether any internal leaks exist in the engine. Such tests determine oil-change intervals, parts wear, and oil contamination.

To summarize—

Lubricating oil

- puts a film between moving parts; this film prevents metal-to-metal contact and reduces friction and wear.
- cools internal engine parts.
- forms a pressure seal between the combustion chamber and the crankcase.
- removes gummy compounds from inside an engine.
- operates the engine's governor.
- operates shutoff controls.

Filters and strainers

- remove dirt from oil; strainers take out large particles, and filters remove smaller particles.
- have bypass valves so that if the filter clogs, unfiltered oil will flow to the engine.

Oil coolers

- keep oil at a temperature below that which would cause the oil to lose its ability to lubricate.
- have a bypass system so that cold oil does not go through the oil cooler but waits until it gets warm.

Oil lubricates

- crankshaft, camshaft, and pushrod guides.
- pistons, piston pins, and piston liner.
- valve mechanisms.
- timing gears.
- fuel-injection system and governor.
- turbocharger and supercharger (blower).

Cooling Systems

▼
▼
▼

s an engine runs, it produces heat. This heat comes from the burning of fuel and from the friction of moving parts. About a third of the heat put out by an engine goes into mechanical energy (work). The engine has to get rid of the other two-thirds, or it will overheat and stop running. Another third (approximately) of this heat is lost through radiation (heat thrown off by the engine) and through the exhaust system. The engine's cooling system takes up (absorbs) the remaining one-third of the heat.

Purpose of the Cooling System

Normal combustion of fuel in an engine produces temperatures as high as 3,000° to 5,000°F (1,600° to 2,800°C). Much of this heat goes to the cylinder heads and walls, the pistons, and the valves. Unless the cooling system carries this heat away, it damages the engine. A cooling system, therefore, prevents damage to vital engine parts. The cooling system also keeps the parts cool enough to work at their best. That is, a part running too hot may not fail, but it will not give maximum performance.

Coolant

The engine transfers heat to a fluid called *coolant*. A common coolant is a mixture of water, ethylene glycol (commonly called antifreeze), and other additives. Antifreeze not only lowers the freezing point of water, it also raises the boiling point. Thus, a mechanic adds antifreeze to the water in the cooling system to keep the water from boiling away as well as to keep it from freezing.

Another additive an engine operator may put in the coolant is a *corrosion inhibitor*. Called chromates, inhibitors delay the formation of rust and cut back on corrosion in the cooling system.

As the engine runs, it circulates the coolant through passages and openings around the cylinders. The engine also drives a *coolant pump* (a water pump). The coolant picks up the heat from around the cylinders. The water pump sends the hot coolant to a radiator or to a heat exchanger. The radiator or the heat exchanger then transfers the heat to the air or to another, cooler liquid. The water pump circulates the cooled-off coolant back into the coolant passages and openings to repeat the heat-removal process.

Radiators

A *radiator* is a bundle of hollow tubes through which the engine circulates coolant (fig. 50). Hot coolant from the engine enters the top of the radiator. It goes into a small tank—a surge tank—on top of the tubes, then flows into the tubes.

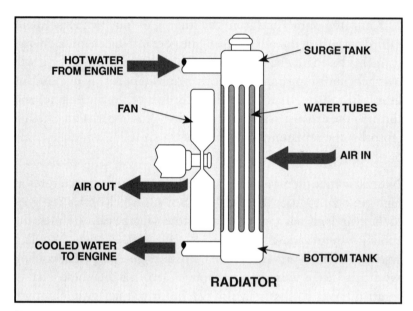

Figure 50. Features of a radiator

Fins

The manufacturer attaches hundreds or thousands (depending on the radiator's size) of very thin, heat-conductive metal *fins* to the outside surface of the tubes. The fins draw heat away from the tubes, which are hot from the coolant, and radiate heat to the air.

Engine Fans

The engine drives a fan that blows air over it and, at the same time, draws air across the radiator's fins and tubes. The blown air removes the heat from the fins. By the time the coolant gets to the radiator's bottom tank, it is much cooler than when it entered the top. The water pump circulates the cooled-off coolant back to the engine to start the cycle over.

A *heat exchanger* is like a radiator except that moving air does not carry away the heat in the coolant. Instead, a liquid from an outside source that is cooler than the hot engine coolant carries the heat away. A heat exchanger, like a radiator, has several tubes. Unlike those of a radiator, heat exchanger tubes do not have fins. The engine circulates coolant through the tubes (fig. 51). The tubes are surrounded by a moving raw water bath that is a lot cooler than the hot coolant coming out of the engine. So, in a heat exchanger, instead of air carrying the heat away, raw water removes the heat from the engine coolant. (Raw water is water taken directly from a source, such as the sea, and it is not treated.)

A heat exchanger has some advantages over a radiator, especially on large, stationary engines (like the ones on a rig). For one thing, the engine does not have to provide power to a fan. For another, the lack of a fan reduces noise. For still another, heat exchangers take up less space than radiators. Compactness is important offshore, where personnel and equipment take up every bit of space. Finally, heat exchangers work better than radiators because the temperature of the water that serves as the cooling element in a heat exchanger is usually a lot cooler than the temperature of the air that cools a radiator.

Heat Exchangers

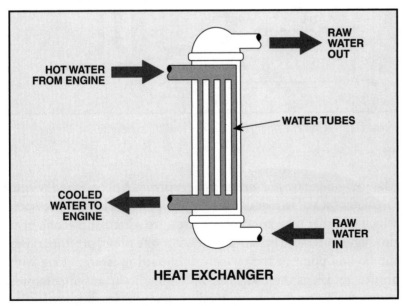

Figure 51. Features of a heat exchanger

Coolant Flow

Whether a rig's engines use radiators or heat exchangers, the key to cooling the engine is the flow of coolant through it (fig. 52). When the engine runs, it drives a coolant pump (the water pump), which moves coolant to the oil cooler. After cooling the oil, coolant goes into the cylinder block, where it flows around the cylinder liners and into the cylinder head. It then flows around the combustion chambers and into pockets (jackets) around the valves. After removing heat from all these places, the coolant goes through pipes (a *manifold*) and into the top tank of the radiator or heat exchanger. The coolant then flows through the radiator or heat exchanger where it is cooled and recirculated through the system.

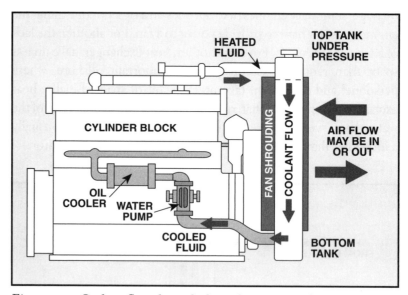

Figure 52. Coolant flow through the cooling system of an engine

Pressurized Cooling

Many engines have *pressurized cooling systems*. A pressurized system circulates the coolant under 5 to 15 psi (35 to 105 kPa) of pressure, which is higher than atmospheric pressure. Circulating coolant at this higher pressure has advantages. For one thing, pressure raises the boiling point of the coolant. Increased pressure, along with antifreeze, keeps the coolant from boiling away at temperatures above 212°F (100°C). Some modern engines circulate coolant at temperatures higher than the boiling point of water; however, most maintain a temperature of 185°F (85°C).

Even though the coolant temperature is above water's boiling point, that temperature is still a lot lower than the heat produced by the engine. As long as engine parts get no hotter than about 800°F (425°C), they can operate well. Coolant temperatures of 225° to 250°F (107° to 121°C) are so much lower than the temperature of the engine's parts that the coolant easily reduces the much hotter temperatures.

Another advantage of a pressurized cooling system is that the coolant cannot evaporate. A pressurized system is closed and therefore does not expose the coolant to the atmosphere, where evaporation can occur.

Finally, because the coolant is hotter than the normal (unpressurized) boiling point of water, it gives up its heat very readily to the surrounding cooler air. For instance, assuming that the temperature of the air in the engine room is 115°F (46°C), the difference in temperature between 250°F (121°C) and 115°F (46°C) is 135°F (75°C). If the coolant was at 185°F (85°C), the difference would only be 70°F (39°C). The larger the temperature difference, the faster the heat transfers.

More About Radiators

Even though almost everyone calls it a radiator, a radiator is really a liquid-to-air heat exchanger. Heat in the liquid coolant goes into the air that is forced through the radiator by a fan. A good flow of cool air through the radiator is absolutely essential to efficient heat transfer. Likewise, a good flow of coolant is also necessary to carry heat away from the engine.

When two fluids (like air and liquid coolant) are in contact at different temperatures, the two temperatures tend to equalize. That is, the hotter fluid warms the cooler fluid and the cooler fluid cools the hotter fluid. So, to cool an engine with a radiator, the air must be at a lower temperature than the coolant, with no less than a 10° to 15°F (5° to 8°C) temperature difference. An 85°F (47°C) difference is better. The smaller the temperature difference, the larger (and thus more expensive) the system has to be to give adequate cooling.

Radiator Size The engine manufacturer must provide a radiator that is able to cool the engine under the most severe conditions. A radiator that is too small may cause the engine to overheat; moreover it restricts the flow of coolant. As a result, the water pump, unable to keep a positive pressure on the restriction created by the small radiator, may develop negative pressure. With negative pressure, the pump draws air into the system. Air in the coolant reduces its cooling ability, since air does not transfer heat as well as liquid does.

Coolant Flow The builder must also provide a radiator that allows the engine to circulate an adequate amount of coolant through it. In fact, the volume of flow through the radiator is just as important as the number of tubes and fins and the depth (height) of the radiator.

Moreover, the manufacturer must install coolant manifolds that do not restrict the flow of coolant to and from the radiator. Crew members should therefore not make any modifications that may restrict coolant flow. Also, when crew members or mechanics install or reinstall coolant hoses or pipes, they should make sure that they do not trap any air at the connections. Air pockets can block coolant flow and let air into the system, which causes corrosion.

Fans Engine makers ensure that the radiator fan is the right size. They design the fan to run at the correct speed to move air efficiently through the radiator. Fan design takes into careful consideration the necessary number of blades and the best spot for optimal operation. Crew members should therefore not make any modifications to the engine fans without consulting the manufacturer.

Thermostats (Temperature Regulators) A temperature regulator (commonly called a *thermostat*) controls the flow of coolant from the engine's cylinder heads to the radiator. (Thermostats control flow in heat exchangers as well.) A thermostat is a temperature-sensitive valve. When the engine coolant is cold, the thermostat closes. When it is closed, coolant bypasses the radiator and flows directly to the inlet side of the water pump and back through the engine. When the engine coolant reaches operating temperature, the thermostat opens, allowing coolant to flow through the radiator.

Running an engine without a temperature regulator may cause damage. The engine may run too cold or too hot. Most often, an engine runs too cold without the regulator. An engine running too cold quickly builds up sludge in the lubrication system. (High temperatures prevent sludge from forming.) Sludge may soon plug the oil filter as well as the oil lines and the passages. Plugged lines and passages keep oil from getting to the bearings and other engine parts, and the result is engine breakdown. Furthermore, an engine running at reduced temperature may use too much oil. At reduced temperatures, the clearance between the cylinders and the piston rings increases. As a result, more oil gets past the rings and the engine wastes the oil.

On the other hand, running some engines without a temperature regulator may make them run too hot. Engines that have a bypass between the temperature regulator and the inlet side of the pump will run hot without the regulator. With the regulator removed, coolant bypasses the radiator and flows directly to the water pump, causing overheating.

How Radiators Work

A radiator is a simple device with no moving parts. It reduces the temperature of the coolant flowing through it. The coolant can then efficiently cool the engine.

Top Tank

Hot coolant from the cylinder heads enters the top (surge) tank of the radiator (see figure 50). The top tank provides space for the coolant to expand into as the engine heats it. It also serves as storage space for coolant and directs it across the tubes (the core) of the radiator. The *radiator core* is the heat exchanger. Along with the fins, it transfers the heat carried by the coolant to the air.

Fins The fins on the tubes pass heat to the air that flows through the radiator by providing large surface areas for exposure to cooling air. Radiator damage such as that shown in figure 53 reduces air flow, while dirt and trash between the fins keep them from radiating heat (fig. 54).

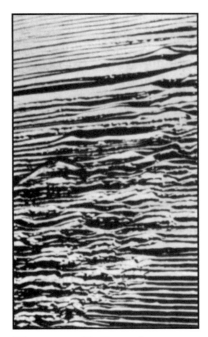 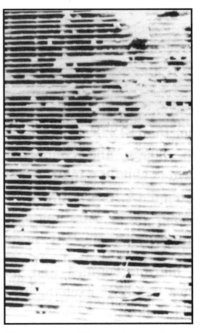

Figure 53. Radiator damage reduces air flow.

Figure 54. Dirt and dust plug air passages in a radiator.

Tube Spacing How the tubes are spaced determines the capacity of the radiator. The manufacturer must put enough tubes in the radiator to allow the coolant to flow freely. Too many tubes, however, block air from the tubes farthest from the front of the radiator. That is, tubes that are behind others may not get very much air flow across them because the front-most tubes block it. The radiator designer has to balance putting in enough tubes to allow free flow of coolant against putting in so many that some do not radiate heat to the air.

Rig builders try to locate the engines in such a way that cooling air passes through the radiator core only once. If hot air—air that has already passed through the radiator—backflows through it, the hot air cannot remove as much heat as cool air. Also, backflow interferes with the flow of cool air, further reducing heat removal. Thus, rig designers often install air tunnels (air ducts) that force flow over the radiator and prevent backflow, ensuring that only cool air reaches the radiator.

Wind can interfere with air flow through the radiator. A strong wind blowing against the direction of flow can create a dead space around the radiator. For example, if the fan is moving air to the south and the wind is also coming from the south, the fan blows against the natural flow of air. As a result, a strong wind can reduce or stop air flow. To prevent air stagnation, crew members can put up wood or metal barriers to block the wind. Or, automatic shutters mounted in front of the radiator can regulate air flow (fig. 55). The shutter control senses when wind direction is offsetting the direction of air flow and partially closes the shutters. The partially closed shutters block the wind but allow air to flow through the radiator.

Air Flow

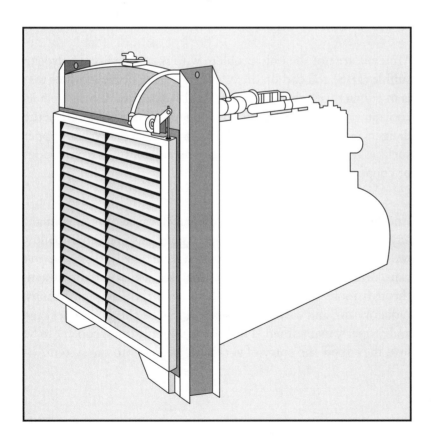

Figure 55. Automatic shutter arrangement for radiator

Coolant Water Quality

The quality of the water used to make up the coolant should be high. Experts say that water too dirty to drink is too dirty to put in a radiator. Foreign material in the water can reduce the ability of the coolant to remove heat.

Dirt

Dirt in the coolant can plug water passages in the engine and large particles can completely block radiator tubes. Blocked tubes prevent coolant from circulating properly and the engine can overheat. It is very important, therefore, to use coolant that does not have dirt, clay, or trash in it.

Dissolved Minerals

Dissolved minerals in the water can also be a problem. If the coolant boils, dissolved minerals form very hard deposits that stick tightly to internal surfaces. Calcium carbonate (lime) is a common deposit, because it is found in a lot of water sources. Lime deposits are very poor conductors of heat; instead, they are good insulators. As a result, a lime deposit of ⅟₃₂ in. (less than 1 mm) can insulate as much as 2 in. (50 mm) of cast iron. Very little heat can radiate out of the engine into the coolant passages if lime deposits exist.

H_2S and CO_2

Minerals are not the only problem with water quality. Hydrogen sulfide (H_2S) and carbon dioxide (CO_2) often occur in water that is in contact with decaying leaves and vegetation. If a rig gets its coolant water from a creek or pond, operators must be aware of the dangers posed by H_2S and CO_2. Both substances corrode copper surfaces, and manufacturers often make radiator tubes from copper or copper alloys.

Oxygen

Another corrosive substance is oxygen. Because oxygen is in air, anywhere air is dissolved in water, oxygen is also present. Shallow wells, creeks, ponds, and the like often have dissolved oxygen that can corrode the cooling system. Air can also enter the system through loose connections. All hose connections, pipe fittings, radiator caps, and water pump seals should be checked for leakage and properly maintained. These leaks not only cause coolant to be lost; they allow the entry of corrosive oxygen into the system.

1. Only clean, soft water that is free from silt should be used. Organic material or sulfur should not be allowed to contaminate it.

2. The engine's head gaskets must be properly tightened. Head gaskets provide a seal not only between the cylinder head and the crankcase, but also between the oil passages and coolant passages in the head and crankcase. Loose and leaky head gaskets can let coolant get into the oil and contaminate it.

3. Water pump seals must be tight and free of leaks; otherwise, air can get into the cooling system.

4. All hose and pipe connections must be tight and free of leaks.

5. The top tank in the radiator must always be full of coolant. A full top tank keeps out air.

6. The water pump must be lubricated according to the manufacturer's specifications for lubricant, procedure, and schedule.

7. Oil coolers must be kept clean and free of sludge and buildups.

8. The cooling system should be cleaned and flushed on a periodic basis, following the instructions of the manufacturer or the rig supervisor.

9. The coolant should be tested regularly according to the specifications in the rig maintenance program. Antifreeze and other additives should be renewed as often as is necessary to keep them working well.

How to Protect the Cooling System

Cooling System Checks

Engine operators should check the cooling system on a regular and routine basis. Some items to check include—

1. the tension on the belt that drives the water pump. This belt runs from an engine pulley to a pump pulley. The belt tension must be kept tight enough to turn the pump at the proper speed. If it is too loose, the belt slips and does not turn the pump properly. A belt with too much tension, however, can wear out the pump bearings very quickly, so it should not be overtightened.

2. the water-pump shaft packing, or seals (fig. 56). These must be checked daily to ensure that they are not leaking. If water can get out, air can get in, and air in the cooling system cuts down on its efficiency.

3. the hoses, the connections, and the radiator pressure cap. These items must be checked daily to ensure that they are airtight. As the bottom (suction) radiator hose becomes softened by use, it can collapse, restricting coolant flow to the water pump. In the same way, a leaking radiator cap must be immediately replaced. In pressurized cooling systems, a leaking cap cannot keep pressure on the system. As a result, coolant could begin to boil.

4. the thermostat. At room temperature it should be fully closed. When placed in water heated to 170° to 190°F (77° to 88°C), the thermostat should open completely.

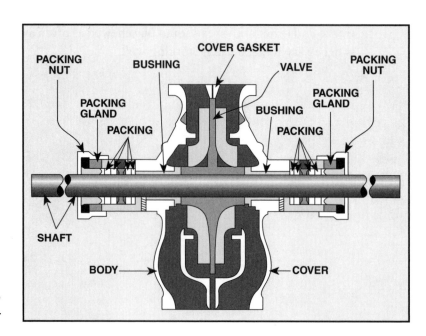

Figure 56. Water pump shaft packing

Rig builders often put the engines in a room or in some kind of enclosure. It is important for them to provide adequate ventilation in an enclosed space. If they do not, an engine can overheat, even if its internal cooling system is working well.

As an example, assume that crew members installed three 350-hp (245-kW) engines in a rig compound. In one minute of running time, the engines burn 159,000 heat units. Of these 159,000 units, the work the engines do accounts for only 44,500 units (about 28%). Subtracting 44,500 from 159,000 leaves 114,500 heat units to account for. The exhaust stack disposes of 47,500 heat units. The engines throw the remaining 67,000 heat units into the air around them. The production of 67,000 heat units per minute could increase the air temperature in the engine room by 420°F (216°C) every minute.

In short, it can get very hot around the engines unless the rig builder provides proper ventilation. A good ventilation system replaces the engine room air as fast as it heats up. In the example, the ventilation system would have to move 186,000 ft³ (5,265 m³) of air every minute, assuming a 20°F (11°C) rise of 60°F (16°C) outside air. The fans on the engine radiators can move this much air if the engine room has enough openings to outside air. If the engines overheat, yet their cooling systems are in good working order, then it may be that there are not enough openings in the room or enclosure. What is more likely, however, is that a crew member may have blocked one or more of the openings. Rig personnel must ensure that all openings are clean and unblocked by any debris or equipment.

Engine Rooms and Ventilation

More About Heat Exchangers

A heat-exchanger cooling system uses a bundle of tubes inside a closed shell (fig. 57). Note that a heat exchanger is small compared to a radiator, even though the whole exchanger system takes up as much space as a radiator, or even more.

A heat exchanger consists of two separate water systems. One is the raw water system. Raw water, instead of air, removes heat from the engine coolant. The second water system is the engine coolant system. This system is exactly like the coolant system in an engine with a radiator.

Raw water, usually sea water if the system is offshore, enters the heat-exchanger shell and circulates around hollow tubes. A separate pumping system moves raw water through the exchanger's shell. Engine coolant circulates through the tubes. Since the raw water is cooler than the engine coolant, the raw water removes heat. Note that coolant and raw water never come into direct contact. Coolant stays inside the tubes while raw water circulates in the shell outside the tubes.

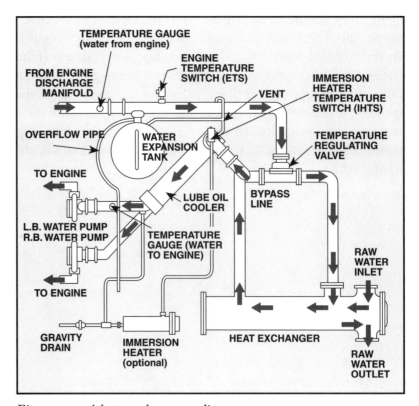

Figure 57. A heat-exchanger cooling system

Raw water exits the shell after removing heat from the coolant in the tubes. A special pump may recirculate the raw water. Offshore, however, where there is an abundant amount of cool sea water available, the system usually does not recirculate it.

Hot coolant from the engine goes through a discharge manifold. Many systems have a temperature gauge at the exchanger. The operator can read the coolant temperature leaving the engine from this gauge. An *engine temperature switch* (ETS) senses overheating and shuts down the engine if overheating occurs.

Coolant then goes through piping and passes a temperature-regulating valve (a thermostat). When the valve is cold, it closes to divert coolant through a bypass line to the oil cooler.

In extremely cold climates, such as in Alaska, Siberia, and the like, an *immersion heater temperature switch* (IHTS) senses very cold coolant and turns on an immersion heater. The heater helps heat the water as the engine warms up.

After leaving the oil cooler, coolant goes through two lines to two water pumps. Each pump feeds coolant to the one side of the engine. A temperature gauge in the inlet allows the operator to check coolant temperature as it enters the engine.

A water-expansion tank completes the installation. As the coolant heats up and expands, it vents into this tank. An overflow pipe on the expansion tank prevents the expansion tank from overfilling.

Note that the rig builders always place the heat exchanger lower than the highest point of the engine because this keeps air out of the system.

To summarize—

The cooling system

- removes about one-third of the heat produced by an engine.
- carries heat away from cylinder heads, pistons, and valves.
- prevents damage by keeping engine parts cool.

Radiators

- transfer coolant heat to the air via tubes, fins, and engine fan.

Heat exchangers

- transfer coolant heat to water via tubes surrounded by the water.

Coolant

- must be kept clean.
- must not have too many dissolved minerals in it.
- should not have any H_2S and/or CO_2 in it.
- must be free of oxygen.

Cooling system checkpoints

- water pump drive-belt tension
- water pump seals
- hoses, connections, radiator cap
- thermostat

Air-Intake Systems

Adiesel engine needs air to run. In each engine cylinder, a piston compresses the air to a high temperature. When the engine injects fuel into the hot air, the fuel-air mixture burns to provide power. A typical diesel needs 12,000 gal (45 m³) of air for every 1 gal (3.8L) of fuel it burns. If the engine does not get enough air, it will overheat, and carbon deposits will rapidly form.

The air for combustion must be clean. Even a small amount of dust in the air going into the engine can grow large in a few hours because of the large amount of air the engine needs to run.

As mentioned earlier, manufacturers naturally aspirate some engines: a *naturally aspirated engine* takes in air at atmospheric pressure. Manufacturers also supercharge some engines: a supercharger forces air into the engine under pressure. Engine builders must supercharge two-stroke engines. The supercharger (the blower) forces exhaust gases out of the cylinder and injects intake air for combustion.

On two-stroke engines, the blower is a gear-driven compressor; that is, the engine drives the compressor by means of gears on the engine that mesh with gears on the compressor. On four-stroke engines, the blower is a high-speed centrifugal compressor—a turbocharger. The engine's exhaust powers the compressor (the turbine). Whether on a two-stroke or a four-stroke engine, the blower always compresses the air and forces it into the intake manifold.

Compressed air is hot. Depending on engine size, the temperature of the intake air can be as high as 350°F (177°C). Air at such high temperatures can make the engine run too hot, resulting in damage to pistons, valves, valve guides, and other parts.

To reduce the temperature of supercharged air, manufacturers equip engines with an *aftercooler*. The aftercooler uses cool water or, sometimes, cool air to lower the temperature of the supercharged air.

Air Cleaners

In both naturally aspirated and supercharged engines, air cleaners remove much of the dust and dirt that is in the air. All of the air that goes into the engine should pass through the cleaners first.

Types of Air Cleaners

Manufacturers make two types of air cleaner: dry and oil bath. A *dry air cleaner* uses centrifugal force and filter elements to remove dust and dirt from the intake air. An *oil-bath cleaner* uses oil with filter elements to remove particles from the air.

A cleaner not only has to filter the air, but it also has to let in the correct volume of air. Because dirty air cleaners reduce the volume of air, it is important to clean or replace air cleaner elements when they become dirty. Note that air cleaners do not filter out every particle of dust. If they did, the operator would have to service them (change or clean the filter) so often as to be impracticable.

No definite rule exists on how often to clean or replace the filtering element of an air cleaner. A cleaner operating under very dusty conditions needs more frequent service than one operating in relatively clean air. Some air cleaners have an indicator that gives a visual sign or warning when the cleaner needs servicing; however, perhaps the best way to set up a good servicing schedule is to inspect the cleaners frequently during normal operation and then set up a schedule that is reasonable for that operation.

Dry Air Cleaners

Dry air cleaners, unlike oil-bath air cleaners, do not use liquids to trap dirt and dust. Paper elements that fit inside a housing, along with the centrifugal action of the air going through the elements, act to clean the air. Paper elements that fit inside a housing, along with the centrifugal action of the air going through the elements, serve to clean the air.

Operation

Heavy-duty dry air cleaners filter the air very well under all operating conditions (fig. 58). Air enters the cleaner through a perforated steel housing, then passes through a disposable safety element that removes the biggest dust particles.

Inside the housing are steel tubes with vanes. After passing through the safety element (a filter), the air crosses the vanes, which cause the air to swirl. This swirling motion creates centrifugal force, throwing the dust particles against the walls of the tubes. Part of the intake air then carries this dust into chambers. Once in the chambers, special "vacuator" valves automatically open to dump the dust accumulated in the chambers. On models without valves, the dust collects in the chambers, and the operator removes it by hand.

The remainder of the intake air reverses direction in the tubes and spirals back along them. As centrifugal force continues to remove dust particles, the tubes make the air reverse direction again and enter a paper filter element. This paper element filters the air once more before it enters the engine. With care, the operator can clean the paper element several times before replacing it. A hinged door on the side of the housing gives access to the paper element, the safety element above it, and a retainer.

Rubber gaskets seal the elements and the housing. The housing also has mounting flanges and outlets for the filtered air. At the bottom of the housing is an inspection cover that can be removed to allow the operator to manually remove dust from the chambers.

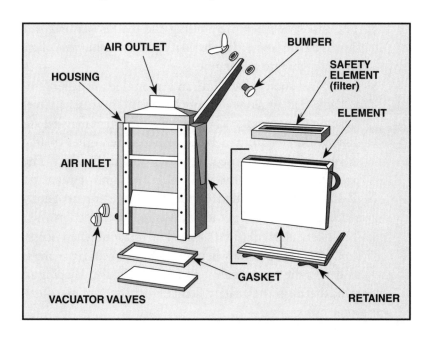

Figure 58. Parts of a heavy-duty dry air cleaner

Cleaning the Air Cleaner

The air cleaner should be serviced according to the manufacturer's recommendations or the schedule worked out for the rig. To clean a dry air cleaner—

1. The tubes, dust chamber, and paper element should be blown out with compressed air at a pressure no higher than 40 psi (276 kPa). Higher pressure can tear the element. Air is directed inside the paper filter element, where most of the dirt collects. The element must be cleaned with care; beating or rapping it can cause damage.

2. If compressed air is not available, a gentle stream of water may be run through the paper filter element; however, a high-pressure stream of water should be avoided since it can damage the element. The element should be cleaned from the inside. If a mechanical element dryer is used, the dryer's air should be circulated and not allowed to get hotter than 180°F (82°C).

3. The cleaned element should be inspected before reinstallation by shining a bright light through it. If light can be seen through it, dust can penetrate it as well. Such an element should be discarded and replaced with a new one.

4. The tubes inside the housing should be examined, using a flashlight, to make sure they are clean. If the tubes are dirty or plugged, they should be cleaned with a stiff fiber brush or compressed air that does not exceed 40 psi or 276 kPa of air pressure.

5. The gasket between the housing and the inspection cover should be checked and replaced if damaged; otherwise, dirty air could get into the engine.

6. Sometimes soot, carbon, lint, and/or oil clog the paper filter so badly that air alone cannot clean it. In this case, a filter-cleaning compound or a nonfoaming detergent may help to dissolve the clogs. But fuel oil, gasoline, or any other volatile solvent should not be used, since they can ruin the filter. The paper filter element may be soaked in a solution of water and filter cleaning compound (or nonfoaming detergent), gently swirling the element in the solution to loosen the dirt.

7. The safety element (filter) should be replaced when necessary. One way to know when this element needs replacing is to determine how much it restricts air intake, keeping in mind that even a new safety element restricts the air intake to some extent.

After the paper filter element has been replaced and the tubes have been cleaned, the degree of air-intake restriction can be measured by checking the pressure of the intake air before it goes through the safety element and comparing it to the pressure after it leaves the cleaner. If the pressure drop across the cleaner is acceptable according to the manufacturer's specifications, then the safety element does not need replacing. If, however, the pressure drop is greater than specified, it is time to replace the safety element. (It is important to remember that this procedure assumes a new or a clean paper filter element. The manufacturer's service manual will give detailed procedures and the equipment required to measure pressure drop in other circumstances.)

Oil-Bath Air Cleaners

Two types of oil-bath cleaner are available: a heavy- and a light-duty model. They both work the same way; this text will focus on the heavy-duty model.

Parts

The housing of a heavy-duty oil-bath air cleaner is a hollow metal cylinder (fig. 59). A relatively small amount of engine oil (the oil bath) lies in the bottom (the sump) of the housing, contained in an inner cup and an outer cup. Two clamps hold the oil cups on the bottom of the housing. A removable screen rests on top of the oil cups, and above that is a fixed, metal-and-wool filter element.

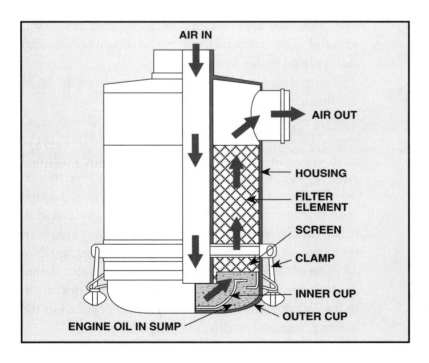

Figure 59. Air flow through a heavy-duty oil-bath air cleaner

Operation

Air enters through an opening in the top and flows down a center tube to the oil bath. The air then goes up through the removable screen assembly. When the air flow changes direction, larger particles of foreign matter slow down and the oil and the screen assembly remove them from the air. The particles then settle in the oil cups at the bottom of the housing.

The air continues upward through the fixed element, which removes the finer particles and the entrained oil. The air goes out a side outlet near the top of the cleaner and flows through the engine's air-inlet housing to the intake side of the blower.

Servicing

To service an oil bath cleaner:

1. The oil cups are removed from the cleaner housing after the clamps that secure them have been loosened. The screen is removed from the cups, and the dirty oil is emptied and properly disposed of. The inner cup and the outer cup are then separated, and both are cleaned with fuel oil. The fuel oil must be properly handled so that it does not harm the environment or personnel.

2. The removable screen may be washed in solvent and dried with compressed air. Its cleanness may be checked by shining a light through it. An even pattern of light should shine through; any blockages are unremoved dirt. If solvent and air cannot clean the screen, it should be discarded and replaced with a new one.

3. The central tube may be cleaned by running a swab soaked in solvent through it.

4. The oil cups are put back together and filled with clean engine oil to the level indicated by a scribed line on the inner cup. The oil used should be of the same grade and viscosity as that in the engine. The oil cups must not be overfilled, since the intake air could pull excess oil through the air cleaner and into the engine. The oil has dirt in it, which could harm the engine. Also, the engine could burn the excess oil and overspeed. On the other hand, too little oil reduces the efficiency of the air cleaner. Once the oil level is correct, the removable screen is placed back on top of the oil cups and the assembled unit is replaced in the housing, securing it with the clamps.

All gaskets and joints should be checked for tightness. All connections from the air cleaner to the engine should be checked for air leaks to prevent air from bypassing the air cleaner.

5. At regular intervals, or whenever the fixed element becomes plugged, the entire air cleaner should be removed from the engine and the fixed element should be cleaned. Taking off the oil cups from the bottom of the cleaner, the operator should pump a suitable solvent (kerosene, but not gasoline) through the air outlet. The solvent must be handled properly so that no harm comes to personnel or the environment. Enough pump force should be applied to produce a hard, even stream of solvent from the bottom of the fixed element. This reverse flushing should continue until the solvent removes all foreign material. (Note: When using a solvent recirculating system, the operator should check to make sure that the tank is big enough to let contaminants settle out before the system recirculates the solvent. In a solvent recirculating system, used solvent goes back into a tank for reuse.) If reverse flushing fails to clean the fixed element, it should not be removed from the housing, but the entire air cleaner should be replaced with a new one.

Additional Cleaner Maintenance

Air cleaners stop working if not installed correctly or if not maintained properly. Problems include leaks in the intake or outlet ducts, loose hose connections, and damaged gaskets. Such problems can let dust-laden air completely bypass the cleaner and directly enter the engine.

To ensure that the cleaner is doing its job, the operator must—

1. keep the cleaner tight on the engine's air intake.
2. properly assemble the cleaner so that all its parts are oil- and air-tight.
3. immediately replace any damaged parts or connections.
4. inspect the cleaner frequently under dusty conditions. On oil-bath cleaners, it is necessary to clean the oil cups often enough to keep the oil from becoming thick with sludge.
5. remove the air inlet housing after servicing the air cleaner, and clean accumulated dirt deposits from the blower screen and the air inlet housing. All intake air passages and the air box must be kept clean.

Blowers (Superchargers)

As mentioned before, engine manufacturers can supercharge (blow) diesel engines. Most two-stroke engines have a blower to force air into the engine. Air going into the engine pushes out (scavenges) exhaust gases from the cylinder after combustion. Manufacturers also blow many four-stroke diesels to increase their power. Some two-stroke engines and many four-stroke engines have a turbocharger. Gases coming out of the engine's exhaust system drive the turbocharger which, like a blower, forces additional intake air into the engine for increased performance. For very high performance diesels, sometimes the engine manufacturer uses both a crankshaft driven blower (supercharger) and exhaust gas driven turbocharger on the two-stroke diesel engine. The atmospheric air is first compressed by the turbocharger and then forced into the supercharger before entering the combustion chamber.

Roots Blowers

Engine builders supercharge modern, high-speed two-stroke engines with a *Roots blower* (figs. 60 and 61). A Roots blower delivers a large amount of air to an engine under as much as 6 psi (41 kPa) of pressure.

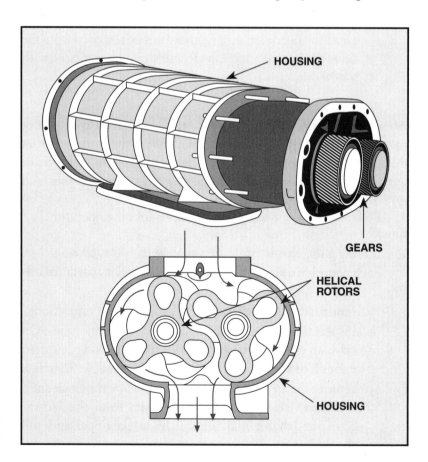

Figure 60. Diagram of a Roots blower

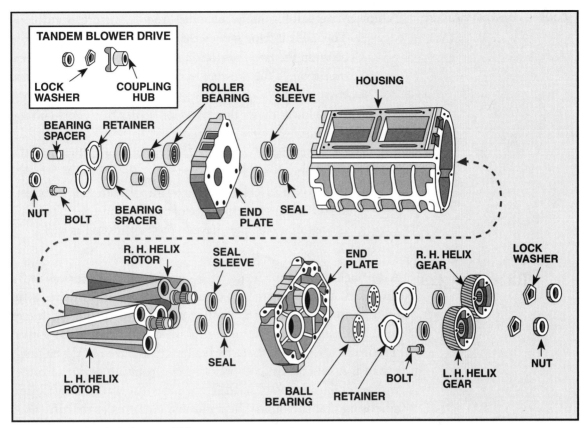

Figure 61. Exploded view of blower components

To compress the air, the blower has two spiraled (helical) rotors, each with blades (lobes) that rotate inside a housing. The engine drives the rotors with gears at about twice the engine's speed, causing them to rotate in opposite directions at the same speed. As they rotate, the lobes compress air drawn through an air cleaner on top of the housing. The compressed air exits the blower from the bottom or the side of the housing and goes into the engine's air-intake manifold.

Each rotor has a shaft at both ends that turns on ball and roller bearings. Seal rings and retainers fit the rotor shafts into an end plate on the front and back of the housing. Intermeshing gears on the end of each rotor transfer engine power to the rotors.

The rotors' helical shape provides continuous and uniform displacement of air. The space (clearance) between the rotor lobes and the housing is close—0.007 in. (0.18 mm) on the leading edge of the lobe and 0.013 in. (0.33 mm) on the trailing edge of the lobe. Close clearances ensure that the blower develops a maximum amount of pressure.

Causes of Blower Failure   A blower may fail for one or more of the following reasons:

1. The intermeshing gears wear and cause improper clearance between the two interlocking rotor lobes. Increased clearance reduces the pressure of the air the rotors supply to the engine.

2. The blower draws in dirt, which scores the rotors and the housing.

3. Loose shafts or worn bearings cause contact between the rotors and the end plates.

4. Worn oil seals let lubricating oil into the housing. Lubricating oil reduces the efficiency of the lobes; instead of pushing against air alone, they have to push against oil as well.

Turbochargers

A turbocharger is made up of a centrifugal compressor and a turbine that work together to compress air for the engine to use in combustion (fig. 62). It is similar to a blower in that it forces precompressed atmospheric air into the engine to increase power. The main difference is that a turbocharger is driven by the engine's exhaust. A centrifugal compressor has a rotating device (a rotor) with several blades on it. As the rotor turns, the blades draw air into a housing that surrounds the rotor and compresses the air.

Located close to the compressor, and sharing the same drive shaft, is a *turbine*. The turbine, like the compressor, has several blades inside a housing. Exhaust gases leaving the engine strike the blades of the turbine to turn it. Since the compressor and the turbine have the same drive shaft, the turbine's movement turns the compressor. As the compressor blades turn, they draw in outside air and compress it. The compressor then forces the compressed air into the engine's air-intake manifold. The pressure of the air entering the manifold depends on the engine's load and the turbocharger's speed.

Modern turbochargers run at 60,000 to 90,000 rpm. This very high speed requires a constant flow of lubricating oil from the oil pump. Moreover, the compressor's housing may get as hot as 350°F (175°C). The turbine housing, because hot exhaust gases drive it, can get as hot as 1,200°F (650°C). Therefore, the compressor and turbine need a lot of oil to cool them. Oil also lubricates the drive shaft and bearings.

Manufacturers also have to consider that turbochargers, because they run at such high speed, take time to slow down after the operator reduces the engine's throttle setting. The engine slows immediately, as does the oil pump's speed and output pressure.

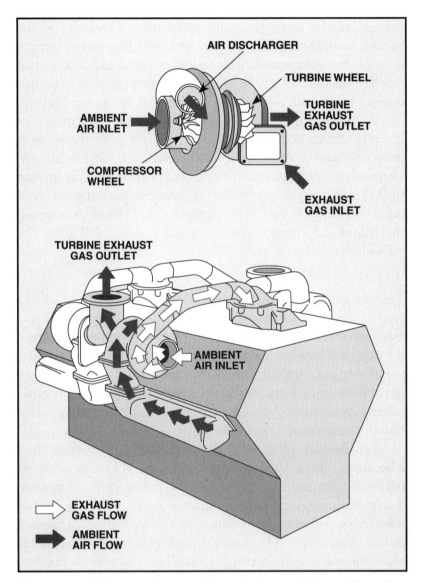

Figure 62. Schematic diagram of a turbocharger (top) and flow of air and exhaust through it (bottom)

The turbocharger, however, in spite of the oil pump's reduced pressure, still needs plenty of lubricating oil, since it continues to turn at high speed for quite a while. Manufacturers therefore modify conventional oil pumps and the lubrication system to ensure that the turbocharger continues to get enough oil during engine-speed reductions.

Aftercoolers

Compressed air leaving the turbocharger is hot. Hot air is undesirable for two reasons: first, it can overheat the engine, and second, hot air is not as dense as cool air; that is, hot air does not contain as much oxygen as cool air. As a result, cool air with more oxygen in it mixes better with the fuel and ensures that all the fuel burns.

To reduce high temperatures, engine builders install an *aftercooler* on turbocharged engines. Coolant from the radiator cools one type of aftercooler; another type is cooled by air. On aftercoolers using radiator coolant, the coolant circulates around the hot air leaving the turbocharger, bringing the temperature of the air very close to that of the coolant. On aftercoolers using air, the engine fan draws air through the radiator and cools the turbocharger's discharged air.

Back-Pressure and Temperature Effects

Turbochargers increase the back-pressure on the engine's exhaust system. They also increase the temperature of the intake air. *Back-pressure* is pressure acting against the free flow of exhaust gases from the engine. Too much back-pressure reduces the engine's power; the pistons and cylinders have to work harder to push exhaust gases out of the engine. Turbochargers increase back-pressure; because exhaust gases are needed to drive the turbine, their free movement into the atmosphere is restricted.

Also, the higher the intake-air temperature rises, the less dense it becomes—that is, the less oxygen it has in it. When the intake air has less oxygen, the engine generates less power from combustion. Turbochargers raise the temperature of intake air because the hot exhaust gases driving the turbine also heat up the turbocharger's nearby compressor. The power gained by supercharging an engine, however, significantly offsets the losses from back-pressure and high temperatures.

To summarize—

To clean intake air

- Two types of air cleaners are used: dry and oil-bath.

Dry air cleaners have

- a perforated steel housing.
- steel tubes with vanes that create centrifugal force to cause dust to fall into chambers.
- "vacuator" valves to remove dust automatically from chambers (some must be manually cleaned).
- a paper filter element, which can be carefully cleaned with compressed air, water, or solvent.
- a safety element, which must be replaced if clogged.
- gaskets, which must make a tight seal; if they fail, they are replaced.

Dry air cleaners should be serviced on a regular basis.

Oil-bath air cleaners use

- an oil bath to which dust adheres.
- a removable screen, which must be cleaned on a regular basis.
- a fixed element, which can sometimes be cleaned but must be replaced if it cannot be adequately cleaned.

Forced-air induction increases the pressure (and density) of the intake air. Two types of forced-air induction are

- superchargers (blowers)—Roots blowers use two helical rotors driven by the engine to compress the intake air.
- turbochargers—turbochargers have an engine exhaust-driven turbine that drives a centrifugal compressor at very high speeds; the compressor increases the pressure and density of the intake air. Turbocharged engines may also use an aftercooler. Turbochargers can be used with either a two-stroke or a four-stroke engine.

Aftercoolers use engine coolant or air to reduce the temperature of the compressed air entering the engine

Exhaust System

The exhaust system's main job is to conduct exhaust gases from the engine cylinders to the atmosphere. A good system takes out exhaust gases with little resistance. If the exhaust system puts too much resistance on the engine's exhaust, back-pressure builds up. As stated earlier, back-pressure is pressure acting against the free flow of gases from the engine. It reduces engine power.

Purposes

Besides conducting exhaust gases from the engine, the *exhaust system* is also designed to reduce (muffle) the noise of escaping gases as they leave the engine. Moreover, the system carries exhaust gases and smoke away from the engines, which the rig builders often place in an enclosed space. And, as mentioned earlier, the exhaust system may power a turbocharger.

Because exhaust gases are so hot, they ignite any flammable foreign material that may be in them. The particle is usually very small, so that it burns as a spark rather than as a flame. Exhaust systems include mechanisms to quench these sparks and remove them from the exhaust gases.

Finally, the exhaust system may also furnish heat to (1) make steam for heating and cleaning the rig, (2) distill seawater to make fresh water, and (3) warm other equipment.

Exhaust System Parts

Except on turbocharged engines, which may not have mufflers, a basic exhaust system (fig. 63) consists of:

- an exhaust manifold
- an exhaust pipe
- a muffler (exhaust silencer)
- a tail pipe

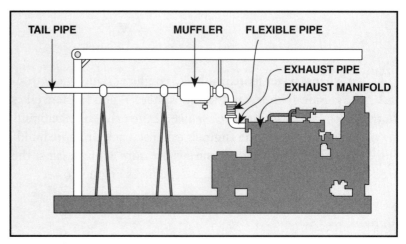

Figure 63. Diesel engine exhaust system

Exhaust Manifold

An *exhaust manifold* connects the exhaust port of each cylinder to a single exhaust pipe. On engines with two banks of cylinders, some manufacturers may install one exhaust manifold to carry the exhaust from both banks, while others may install two manifolds—one on each bank.

On large engines, manufacturers weld all parts of the manifold together and put a *water jacket* around the whole thing. The cooling system circulates coolant through the jacket, reducing the temperature of the manifold. Water-cooled exhaust manifolds lessen the danger of burning those who work around the engine. They also reduce the chance of a fire inside an engine room. What is more, cool exhaust gases in the exhaust manifold reduce back-pressure on the engine. We know that hot air expands, whereas cool air contracts. Expanding hot air puts back-pressure on the engine exhaust, reducing the engine's power. Contracting cool air reduces back-pressure and does not decrease engine power.

The temperature of an uncooled exhaust manifold varies as the engine load varies. The higher the engine load, the hotter the manifold becomes; conversely, the lighter the load, the cooler it becomes. The manifold expands when its temperature goes up and contracts when its temperature goes down. Engine builders therefore use flexible connections to attach the manifold to the engine.

Manifold designers put covered openings in water-jacketed manifolds. By opening the covers, a mechanic or operator can clean out scale that collects on the walls of the manifold. Scale is made up of minerals in the coolant water that adhere to the inside of the manifold.

Exhaust Pipe

The *exhaust pipe* connects the exhaust manifold's outlet to the muffler. The pipe should be short and have as few turns as possible. Manufacturers make exhaust pipe from flexible materials that reduce the transmission of engine vibrations, lessen stresses set up by expansion of hot pipe, and make it easy to install the muffler. Either a part or all of the pipe may be flexible.

When setting up rig engines, the builders may install one exhaust pipe for each engine or one exhaust pipe for each bank of cylinders on each engine. In either case, the pipes that are inside buildings or engine rooms are usually insulated. Insulation keeps the room from getting too hot from the exhaust-pipe heat.

The diameter of the exhaust pipe depends on the horsepower (kilowatts) of the engine. Figure 64 is a graph that shows exhaust-pipe diameters for engines of various horsepower (kilowatts).

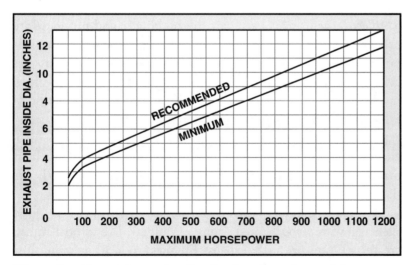

Figure 64. Approximate sizes for diesel engine exhaust piping

It shows two curves: one is the recommended diameter and the other is the minimum diameter. For example, in a 1,000-hp (700-kW) engine, the minimum diameter of the exhaust pipe is 10 in. (254.0 mm). The recommended diameter is 11 in. (279.4 mm). The figures on the graph assume that the exhaust pipe is no longer than 50 ft (15 m) and that it has no more than one right-angle turn. The recommended diameters put very little back-pressure on the muffler: 7 in. (175 mm) of water to be exact.

Like pounds per square inch or kilopascals, inches or millimetres of water is a pressure measurement. Technicians or operators use a manometer to make inch or millimetre measurements. A *manometer* is a U-tube with water in it. Pressure on one side of the U-tube raises the water height in the other side a given number of inches or millimetres, depending on the pressure. Inches (millimetres) of water is useful for measuring small amounts of pressure. One in. of water equals 0.0361 psi; 175 mm of water equals 0.249 kPa. Thus, 7 in. of water is about 0.25 psi; 175 mm of water equals about 43.58 kPa.

An engine's exhaust system can accumulate moisture, particularly throughout long exhaust pipes. Therefore, engine installers often place a *condensate trap* and a drain at a low point ahead of the engine manifold. Operators should periodically check the trap and drain any condensed water. To keep rain or snow out of a vertical exhaust pipe, installers either put a rain cap on the end of the pipe or curve it into a right-angle.

Mufflers

Mufflers (exhaust silencers) quiet the barking sound produced by exhaust gases exiting through the exhaust pipe of naturally aspirated engines. Manufacturers make many kinds. One is a steel cylinder with baffle plates. The *baffle plates*, flat steel sheets welded inside the cylindrical body of the muffler, change the direction of exhaust gas flow. Changing the direction of flow allows gradual expansion of the gases, which is quieter than rapid expansion. Also, covering a muffler with fiberglass or another sound-absorbent material diminishes the ringing sound of the exhaust. It also keeps the muffler from radiating a lot of heat.

Engine assemblers may choose not to install mufflers (noise silencers) with turbocharged engines. Mufflers, like turbochargers, create back-pressure on the exhaust. Engines produce a pulsating flow of exhaust gases. As the pistons move up and down in the cylinders, they push out exhaust gases in rapid bursts (pulses). In a naturally aspirated engine, these pulses go directly into the exhaust system. The pulses create noise, and engine builders install mufflers to reduce it.

On the other hand, in turbocharged engines, the exhaust gases drive a turbine by flowing over the turbine blades to turn them. Because the exhaust gases spread out within the turbine as they turn the blades, the pulses fade out. As a result, exhaust-gas turbochargers put out an almost continuous flow of exhaust gas. Because the exhaust of turbocharged engines is essentially continuous, their exhaust is quieter and may not require mufflers.

A *tail pipe* comes out of the muffler or the exhaust pipe and carries the exhaust gases to the atmosphere. The length of the tail pipe depends on how far it has to conduct the gases from the engine room or the engine enclosure.

Tail Pipe

It is important to remember, however, that engine builders make the tail pipe a certain length to accommodate the way a particular engine expels exhaust gases. Exhaust gases leave the muffler or the exhaust pipe in pulsating waves. When each wave reaches the tail pipe outlet, the atmosphere at the outlet reflects some of wave back down the pipe. The reflected wave should get back to the muffler or the exhaust pipe at the right time. That is, the reflected wave should arrive, not at the moment of a fresh wave, but between two waves. The reflected wave's reaching the muffler or exhaust pipe between new waves prevents back-pressure. Engine installers therefore cut tail pipes to a specific length to make the reflected waves arrive at the muffler or the exhaust pipe at the right time to prevent back-pressure from pulsations. Operators and mechanics should not change a tail pipe's length without first consulting with the manufacturer.

Engine installers cut the end of the tail pipe at an angle other than 90°—say at 45° or so. In other words, they cut it on a bias. Cutting the end of the pipe on a bias reduces noise. If the gases left the pipe in a straight path, as they would from a straight-cut pipe, they would come out with a fast, noisy puff. But when the gases leave on the angled path created by the bias, they gradually spread out and thus make less noise.

When a tail pipe is erected vertically, it is called a *stack* (fig. 65). Where a stack goes through a roof or ceiling, the builders install a flexible collar between the pipe and the opening in the roof. The collar allows the pipe to expand and contract with changes in temperature.

Figure 65. Exhaust stacks

To summarize—

The exhaust system
- conducts exhaust gases from an engine's cylinders to the atmosphere.
- muffles the noise that escaping exhaust gases make.
- carries exhaust gases and smoke away from the engine.
- may power turbochargers.

Exhaust system components
- exhaust manifold
- exhaust pipe
- muffler (exhaust silencer); not always used on turbocharged engines
- tail pipe

Exhaust manifolds
- often have a water jacket for cooling.
- expand and contract with temperature changes.

Exhaust pipes
- connect manifold's outlet to muffler.
- often have condensate drains.

The diameter of an exhaust pipe depends on the engine's power.

Mufflers
- quiet the barking sound of exhaust gases.
- are often insulated.
- may not be used on turbocharged engines.

Tail pipes
- carry exhaust gases from muffler to atmosphere.
- must be the correct length to prevent back-pressure.
- have ends cut at an angle to reduce noise.

Starting Systems

Engine builders install various devices to start diesel engines. Usually, the builder mounts the starter on or near the engine. The *starter* rotates the crankshaft so that the pistons can compress the air in the cylinders. When the air reaches the proper pressure and temperature, it ignites the injected fuel and the engine begins running on its own.

Types of Starters

Electric motors, air motors, hydraulic motors, and gasoline engines are all mechanisms used to start an engine. Another means is to inject compressed air into the cylinders. Very small diesel engines can be started by turning a hand crank.

Why Diesel Engines May Not Start

A diesel engine has to reach a certain speed to build up enough pressure and temperature for ignition. The speed depends on the type and size of the engine, its condition, and the temperature of the surrounding air. In some engines, the starting speed may be as low as 70 rpm, but a small engine may have to be turned up to 3,000 rpm.

Low-Pressure Problems

To ensure a quick start, the starting device must turn the crankshaft fast enough to compress the air to the proper temperature. If the starter turns the engine over too slowly, the air in the cylinders cannot compress and get hot enough to ignite the fuel. Air leaks constantly through the small spaces between the piston rings and the cylinder wall, but if the crankshaft turns fast enough, its speed overcomes the leaks and adequate compression occurs.

Low-Temperature Problems

Heat loss to the cold metal walls of the cylinder may lower the air temperature so much that ignition temperature is hard to reach. Further, badly worn cylinder walls and piston rings allow too much air to leak past them. No matter how fast the crankshaft turns, the air cannot compress enough to reach ignition temperature.

Diesel engines start and run at their best if the average temperature is about 70°F (20°C). An engine that has been shut down for several hours when the temperature is well below 70°F (20°C) may not start promptly. The starter may not be able to turn the engine fast enough to get the necessary pressure and temperature for compression. Or, even if the starter turns the engine over rapidly, the compression temperature may still be too low for ignition.

Lubricating oil gets thicker (its viscosity increases) as the temperature drops. Increased viscosity makes it hard for the starter to turn the engine over quickly. In some installations, a special heater raises the temperature of the engine coolant, which in turn raises the temperature of the oil. Manufacturers equip some small diesel engines with an electric heater in each cylinder's water jacket to warm the oil. In any case, raising the oil temperature reduces its viscosity, thus making engine start-up easier.

To see how low temperatures affect start-up, let us suppose that the intake-air temperature is 70°F (20°C) and that an engine has a compression ratio of 14 to 1 (the piston compresses the air to a pressure 14 times higher than its beginning intake pressure). In this case, the final compression temperature is 700°F (370°C), which is hot enough to ignite the fuel. But if the intake-air temperature of the same engine were to drop to 40°F (5°C), the final compression temperature would be only 200°F (90°C). This temperature is much too low for ignition to occur with normal fuel. Note that most diesel engines have compression ratios much higher than 14 to 1; but, even so, if the intake air's temperature is low, the engine cannot start.

Rig owners install most large engines in rooms where the temperature rarely falls below 70°F (20°C). In situations of greater exposure to the elements, however, the following techniques may be used to raise the temperature of the intake air:

1. using electric intake-air heaters;
2. heating the jacket coolant; or
3. installing glow plugs in the cylinders.

Electric Starters

Rig owners sometimes install *electric starters* on relatively small diesels, such as those used to generate electricity on the rig. Usually, power to turn the starter comes from a *storage battery*, similar to the one in an automobile. The manufacturer sizes the battery to match the engine's cold-weather starting requirements. Cold reduces the battery's output; yet, a cold engine needs a lot of torque to start. Thus, a battery big enough to start the engine in warm weather could fail in cold weather.

Electric starters use *direct current* (DC) rather than *alternating current* (AC). In a DC system, electricity flows in one direction from the battery through wires (cables) to the starter. In an AC system, like the electrical system in your home, electricity flows rapidly back and forth (alternates) between the source and the item being powered. Direct current is better for engine start-up systems because a battery can store direct current for a relatively long time, and the starter can draw on it when needed.

An electric start-up system consists of a storage battery, cables from the battery to a DC motor, and a DC motor. It also has a mechanical connection (usually gears) between the DC motor and the engine crankshaft. The system also has an engine-driven generator to charge the battery. Figure 66 shows an electric starter on a diesel that drives an auxiliary power unit on a rig.

Electric starter motors are heavy-duty devices. They can, however, overheat if anyone operates them for too long a period.

Figure 66. Battery-operated electric starter on a 300-kW auxiliary power unit

To prevent overheating, manufacturers equip some starter motors with a timer that automatically disconnects the power after about 15 seconds if the engine does not start. A motor without a timer should never be operated for more than 30 seconds at a time.

How Electric Starters Work

Two gears connect the starter motor to the engine crankshaft. One is a pinion gear on the end of the shaft of the starter motor (fig. 67). The other is a ring gear, which is on the outer circumference of the engine's flywheel (not shown in the figure). (The flywheel is connected to the crankshaft.) When the operator activates the starter, battery current moves a friction clutch (often called a *Bendix*; see fig. 67) forward.

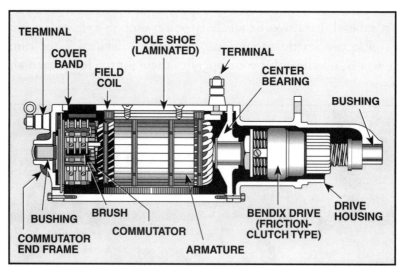

Figure 67. Cross section of a Bendix-type starting motor for small engines

The clutch's forward movement causes the pinion gear to engage the ring gear. The starter motor then begins turning. Since the motor's pinion engages the ring gear, the ring gear turns the crankshaft. If all goes well, compression quickly builds up and the engine starts.

As the engine reaches operating speed, the flywheel's ring gear moves faster than the starter motor's pinion gear. This faster movement causes the starter clutch to move back and disengage the pinion from the ring gear. The starter motor then stops.

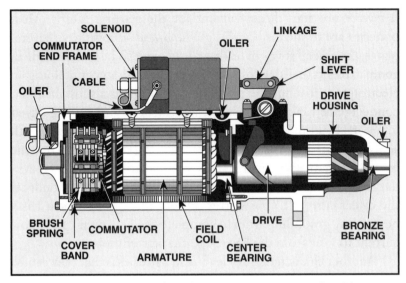

Figure 68. Cross section of an electric starter using a solenoid

Electric starters on large engines use a *solenoid* instead of a friction clutch to engage the pinion gear to the ring gear (fig. 68). (Large engines require a lot of force to engage the pinion with the ring gear. A friction clutch is not up to the job.) A solenoid is a strong electromagnet. When electric current flows to the solenoid, the current causes a shaft to move forward. Linkage between the solenoid shaft and the pinion moves the pinion forward to engage the ring gear. With the gears engaged, the solenoid then completes an electric circuit to turn on the starting motor.

When the engine runs, it turns a constant-voltage generator that recharges the battery. A constant-voltage generator is a generator that puts out the same voltage regardless of the engine's speed, with two limitations. If the engine stops or runs very slowly, the generator cannot charge the battery. To keep the battery from discharging while the engine is idling or stopped, the manufacturer installs an automatic cutout in the system. The cutout electronically disconnects the battery from the generator so that the generator cannot discharge the battery when the engine is running slowly or not at all.

Generators and Cutouts

Batteries

A *battery* provides direct current for the starting motor. Most batteries are of the lead-acid type. *Lead-acid batteries* have several plates, thin metal strips of lead and lead oxide. The plates reside in compartments (cells) within the battery case. Water containing an electrolyte surrounds the plates. (An *electrolyte* is a substance that conducts electricity. In this case, the electrolyte is an acid dissolved in water.) Lead and lead oxide produce electricity when water with acid in it surrounds them. The electricity (DC) flows from the battery's negative terminal (or post) to its positive terminal. Heavy-duty electric wire (cable), connected to the positive post, conducts the direct current to the starter motor. The negative post has a heavy-duty grounding strap to complete the DC circuit and allow current to flow from the battery to the starter motor.

Another characteristic of lead-acid batteries is that they can be recharged. An engine generator can charge them constantly as the engine runs. Or, if the battery loses its charge, an operator can attach a battery charger to the battery terminals and recharge it.

Care of Batteries

Operators should not allow anything that conducts electricity to lie across the two posts (terminals) of the battery. Wrenches, pry bars, or any metal object placed between the two battery posts can short the battery and ruin it. Further, in cold weather, the water in the battery must not be allowed to freeze. If it freezes, the battery will not work.

When a battery is being charged, no sparks or flames can be near it. Each cell of the battery produces oxygen and hydrogen gases during the charging process. Oxygen and hydrogen form a very explosive mixture, which the smallest spark or flame can set off.

One way to check a battery's charge is to measure the specific gravity of the battery's acid solution in the cells. The acid-water mixture's specific gravity goes down as the battery loses its charge. Normally, operators use a *hydrometer* for this check. A hydrometer is a transparent tube with a sphere in it. The sphere indicates the specific gravity of the acid-water solution. The operator draws the solution from the battery into the tube to a specified level, then checks the position of the sphere against a scale drawn on the side of the tube. A fully charged battery gives a hydrometer reading of about 1.260. When the hydrometer reading drops to about 1.250, the operator should recharge the battery.

When connecting a battery charger to the battery, be sure to connect the positive lead of the charger to the battery's positive terminal. Likewise, be sure to connect the negative lead of the charger to the battery's negative terminal. Maintain the charge until the hydrometer reading shows a fully charged battery. Also, if necessary, add distilled water to the battery until the water covers the plates by about ½ in. (13 mm). In hot climates, do not add cold water to a hot battery. The cold water causes the plates to separate and ruins them.

Air-Motor Starters

Air-motor starters work much like electric starters. Compressed air, instead of electricity, operates the starter. Like an electric starter, an air starter has a pinion that engages the teeth on the flywheel (fig. 69). Inside the starter housing, the manufacturer mounts a cylinder equipped with vanes on a rotating shaft. To start the engine, the operator opens a valve to let compressed air into the starter housing, where it strikes the vanes. Because the vanes are at a right angle to the centerline of the shaft, the air strikes them at full force and forcefully turns the cylinder and attached shaft. The turning force of the shaft actuates a starting motor. As it begins to rotate, the starting motor causes the starting clutch to move the pinion so that its teeth mesh with the teeth on the engine flywheel. The turning pinion thus turns the flywheel and the crankshaft. When the operator shuts off air to the starting motor, the pinion teeth disengage from the flywheel teeth. If the starter fails to disengage properly, it can seriously damage the engine.

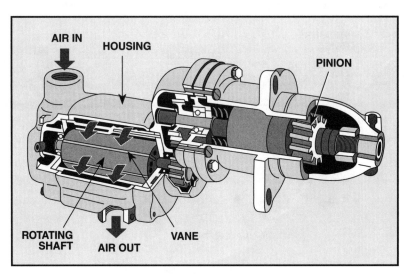

Figure 69. Air-motor starter

Air Requirements

Clean, water-free, and lightly oiled air operates the starter. The pressure for the inlet air should be from 100 psi (700 kPa) to 150 psi (1,000 kPa). An air reservoir that is at least 23 ft3 (0.7 m3) in size is required for a 1,000 hp (750 kW) engine. To remove moisture from the air, a water trap is placed in the air supply line. An oiler installed next to the starter puts oil into the air. Lightweight oil should be used in the oiler so that it can break the oil into a fine mist.

Cleaning an Air Starter

After an air starter has been in service for some time, it can gum up and not work properly. Fine dirt particles in the air and oil mixture build up on the parts of the starter and cause it to malfunction or fail. If the gumming problem is not too serious, the operator may be able to clean the starter without disassembling it. Clean diesel fuel mixed into the air supply may be run through the starter; this procedure sometimes cleans the starter by dissolving the gum.

Hydraulic Starters

A *hydraulic starter* can be used with a diesel engine. The *GM Hydrostarter*™ (fig. 70) is a complete hydraulic system. It stores hydraulic fluid in an accumulator. Also in the accumulator is nitrogen gas. The gas pressures up the fluid to 3,350 psi (23,000 kPa). When the engine operator opens a valve to make the starter engage the flywheel's ring gear, the high-pressure nitrogen forces hydraulic fluid into the starting motor. The motor rapidly accelerates the engine to high cranking speed.

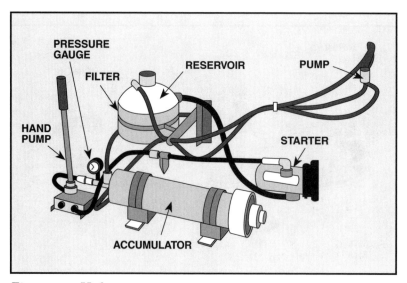

Figure 70. Hydrostarter

The starter system recycles the used fluid, which leaves the starter and goes into a reservoir. An engine-driven pump continuously charges the accumulator during engine operation. If the operator shuts down the engine for a long period, the accumulator loses its charge. The manufacturer therefore provides a hand pump that the operator can use to raise the accumulator pressure to the 1,500 to 2,500 psi (10,350 to 17,250 kPa) required to start the engine.

Operators often install hydraulic starters on engines that generate emergency electricity. Hydraulic starters take very little maintenance and are safer than electric systems. The only real disadvantage of a hydraulic starter is its high initial cost, which is several times higher than that of an air starter.

In some installations, the engine operator uses a gasoline engine to operate a starter motor. The *gasoline engine starter motor* engages the flywheel of the diesel engine. A friction clutch, in combination with V-belts or gears, connects the starter motor to the diesel engine.

The operator starts the gasoline engine by hand cranking it. When it is running, the operator then engages the friction clutch by moving a hand lever. The clutch makes the starter motor's pinion engage the diesel engine's flywheel. When the diesel gets under way, the speed of the flywheel releases the pinion gear on the starter motor.

Operators sometimes use gasoline engine starting systems in very cold climates. They first start the gasoline engine and allow it to run until it is warm. They then route the gasoline engine's warm coolant to the diesel engine. The warm coolant preheats the diesel and makes it easier to start.

Gasoline Engine Starters

Compressed-Air Starting

Large stationary and marine diesel engines often use *compressed-air starters*, in which compressed air is injected directly into the engine's cylinders. A special valve directs compressed air to each cylinder during the power stroke and during the exhaust stroke. The compressed air causes the pistons to move rapidly until enough pressure builds up to ignite the fuel.

The amount of compressed air needed to start a small engine is about twenty-five times its total piston displacement. For a large engine, the amount may only be about ten times its piston displacement. An air compressor pressures up the air anywhere from 125 to 250 psi (850 to 1,750 kPa).

Once a diesel engine has been started, it usually runs for long periods. Because of this, the air compressor is usually fairly small and does not need much power. While the engine runs, the compressor has plenty of time to restore the volume of air required for the next start-up. In some cases, the engine drives the compressor. In others, a gasoline engine or an electric motor drives the compressor. Marine engines often drive compressors, because equipment on the boat uses compressed air for other purposes.

Cylinder-injected compressed air can start a four-stroke engine with five or more cylinders, no matter what position the crankshaft stopped at when the operator last shut down the engine. On a four-stroke engine with four or fewer cylinders, however, the operator must *bar* the engine. That is, the operator must insert a steel bar into a hole on the rim of the engine's flywheel and, using the bar as a lever, manually turn the flywheel to rotate the crankshaft and pistons. The operator must bar the engine until one of the pistons is on the down stroke, slightly past bottom dead center and ready to take starting air.

Compressed air can also start a two-stroke engine with three or more cylinders regardless of the position of the pistons inside the cylinders. If the engine has only one or two cylinders, however, the operator must bar the engine until one of the pistons is on the down stroke.

Regardless of the number of cylinders in a large engine, the operator should bar the engine several times if it has been shut down for quite some time. Barring the engine ensures that the pistons can travel freely in the cylinders. Barring also removes any trapped water above the pistons. Trapped water can prevent the engine from starting.

An operator usually bars a large engine by using a special gear attached to the engine, called a *worm gear*. To make a worm gear, the manufacturer takes a solid metal cylinder and cuts a spiral (helical) groove around it. The worm gear meshes with teeth cut into the engine's flywheel. When the operator actuates the worm gear, it turns the flywheel and the pistons. Usually, an electric or a compressed-air motor operates the worm gear.

To summarize—

The starting system

- rotates the engine's crankshaft so that the pistons can build up enough pressure and temperature to ignite the fuel.
- uses electric motors, air motors, hydraulic motors, gasoline engines, or compressed air.

Electric starters

- use batteries as a source of power
- consist of battery, cables, DC motor, and mechanical connection (gears) between DC motor and engine crankshaft.
- have a pinion gear that meshes with ring gear on flywheel.
- use a battery that is recharged by means of a generator on the engine.

Air-motor starters

- use compressed air to actuate starter.
- have a pinion that engages a ring gear on the flywheel.
- need clean, lightly oiled air at 100 to 125 psi (700 to 1,000 kPa).

Hydraulic starters

- use hydraulic fluid stored under nitrogen pressure.
- use nitrogen pressure that forces high-pressure hydraulic fluid to operate starter motor.

Gasoline engine starters

- use the gasoline engine to rotate the diesel's flywheel.
- operate by means of a manually operated friction clutch that, when engaged, causes the starter's pinion to mesh with the diesel's ring gear on the flywheel.

Compressed-air starters

- are arranged so that compressed air is injected directly into the diesel's cylinders.
- use compressed air to move the pistons to build up ignition pressure.

Instruments

Instruments that keep track of an engine's operation are essential. A doctor needs to monitor a patient's pulse, temperature, respiration, and other vital signs to tell whether the patient is well. Similarly, an engine operator needs to know the engine's temperature, speed, oil pressure, and other signs to tell whether the engine is running well. The vital signs of engines are measured by pyrometers, oil-pressure gauges, oil temperature gauges, coolant temperature gauges, air manifold pressure gauges, and tachometers.

Pyrometers

A *pyrometer*—a thermometer that measures high temperatures—is used to measure the temperature of the engine's exhaust. Alert operators can get a lot of information from pyrometer readings.

Estimating Engine Load

For example, they can estimate an engine's load. Suppose the engine manufacturer recommends that the exhaust temperature not exceed 1,000°F (540°C). The pyrometer, however, shows an exhaust temperature of 1,075°F (580°C). This higher-than-normal temperature indicates that the engine does not have enough horsepower (kilowatts) to adequately power the load the operator has put on it. Overloading an engine wears it out faster than normal; in fact, extreme overload can destroy an engine in a matter of minutes.

Cylinder Temperatures

Each engine cylinder also has a place where the operator can attach a pyrometer and determine the temperature in that cylinder. Should one cylinder read low while the others read high, then the operator knows that the cylinder with the low reading is not firing properly.

On the other hand, should one cylinder read high while the others read low, the cylinder with the high reading is doing too much work; it is overloaded. Overload damages the cylinder.

If the temperature of all the cylinders, except one with a low reading, is normal and the manifold temperature is very high, the operator knows that the cool cylinder is passing raw fuel into the manifold. The abnormally high manifold temperature lets the operator know that the engine is burning the fuel in the manifold. An abnormally high manifold temperature can shorten the life of the turbocharger.

Dividing Loads Equally

In cases where two or more engines drive the same load, pyrometer readings of each engine's exhaust temperature indicate whether one engine is carrying more of the load than the others. The higher the exhaust temperature, the greater the load the engine is carrying. When the operator evenly divides the load between the engines, their exhaust temperatures should be nearly the same.

Oil-Pressure Gauges

An *oil-pressure gauge* shows the pressure on the lube oil system at the point where the operator installed the gauge. From oil-pressure readings, operators can find out a lot about how the engine is running. Either a low-pressure or a high-pressure reading can point the way to problems in the engine.

Low Oil Pressure

If the gauge shows low oil pressure, many things could be wrong. Among these are dilution of the oil, a low level of oil, a blocked suction screen, or a bearing failure.

Oil Dilution

Suppose the oil pressure is normally 50 psi (340 kPa) when the engine is at operating load, speed, and temperature. If the oil pressure gradually drops, diesel fuel may be mixing with the lube oil and diluting it. Diesel fuel can dilute oil when various engine parts wear or fail. For example, worn or improperly installed gaskets or seals can allow fuel to leak past the seal and enter the engine's oil. The oil pump cannot keep the proper pressure on a thin, dilute oil.

Low Oil Level

A low oil level can also cause a low oil-pressure reading. If an oil leak occurs in the engine, then the loss of oil causes the pressure to drop. The oil pump cannot keep the proper pressure if not enough oil is in the engine.

Blocked Suction Screen

Sometimes, a blocked suction screen causes the oil pressure to drop. The pressure gauge on the inlet side of the oil pump should indicate whether the pump is pulling a normal amount of vacuum (pressure below that of the atmosphere) on the inlet. If the inlet vacuum reads higher than normal, and the oil-pressure gauge reads below normal, dirt or trash may be blocking the suction screen. The engine should be stopped and any obstruction in the suction system should be removed.

Bearing Failure

Failure of the engine's bearings also gives a low oil-pressure reading. Bearing failure is serious. If one of the other problems cannot be identified as a cause of low pressure, the operator should stop the engine immediately and get a mechanic to determine whether the bearings have failed.

High oil pressure can also be a sign of trouble in a diesel engine. Oil that is too heavy (too viscous) overloads the oil pump, which causes high oil pressure. Another cause of high oil pressure is a stuck or improperly adjusted regulating valve. Too much oil pressure can wash out bearings, much as a high-pressure water jet erodes a sand bank.

High Oil Pressure

Oil-Temperature Gauges

An *oil-temperature gauge* indicates the temperature of the oil at the point in the system where the gauge's sensor is located. The engine oil temperature should fall within the range set by the manufacturer. Sometimes, the oil-temperature gauge indicates abnormal temperature even though the coolant temperature gauge indicates normal temperature.

Lower-than-Normal Oil Temperature

For example, the coolant temperature may be normal but the oil temperature is below normal. Cooler-than-normal oil is undesirable because it may allow water to condense inside the engine, which could cause damage. What is more, the cool oil causes clearance tolerances inside the engine to vary from normal, and abnormal tolerances can cause premature wear. The lube oil's temperature should be at least as hot as the minimum temperature set by the manufacturer.

Higher-than-Normal Oil Temperature

Two gauges indicate higher-than-normal operating temperature: the coolant-temperature gauge and the oil-temperature gauge. A high oil-temperature reading is a better indication of abnormally high engine temperature than is a high coolant-temperature reading. If allowed to run too hot, the oil breaks down and fails to lubricate properly. Further, oil cools the pistons in a diesel engine, and if the oil is too hot, it cannot perform this vital function.

Manufacturers give maximum oil temperatures for their engines. Normally, however, the oil temperature should be about 10°F (5°C) warmer than the coolant temperature at normal load. Moreover, temperatures above 250°F (120°C) break down the additives in the oil, which causes the oil not to lubricate as well as it should. If the engine overheats while it is under load, the operator should bring both the oil and the coolant temperatures down to normal before putting the engine back under load.

Coolant-Temperature Gauges

Coolant-temperature gauges indicate the operating temperature of the engine. Gauges on the inlet side and the outlet side allow the operator to compare the temperature of the coolant entering the engine with the temperature going out. As a general guide, the inlet temperature should not be more than about 75°F (25°C) cooler than the outlet temperature. Inlet and outlet temperatures should be fairly close, because if cold coolant were to be pumped into the hot engine, the thermal shock to the metal might crack it.

The outlet temperature of the coolant indicates how much heat has entered the coolant from friction and burning fuel in the engine, whereas its inlet temperature indicates how much heat the coolant removed. It is important for the operator to install a temperature regulator so that the cooling system removes only the necessary amount of heat. The water should not be cold.

Inlet Versus Outlet Temperature

All supercharged engines have pressure on the air-intake manifold. A *manifold-pressure gauge* reads this pressure and indicates it on a gauge. Under load and at operating speed, manifold pressure is above atmospheric pressure. The supercharger raises the pressure of the air going into the engine above that of the atmosphere. The engine operator should look at the engine manual to determine what the pressure should read.

Air Manifold-Pressure Gauge

Dirty air filters restrict the flow of inlet air, which shows up as a drop in manifold pressure. The operator should therefore check and clean the air filters. If, however, the pressure does not return to normal, the turbocharger may be malfunctioning. If the turbocharger is not working properly, the operator should immediately notify the engine mechanic to avoid damage to the engine.

Dirty Air Filters

A *tachometer* is an rpm instrument. It indicates the speed at which the engine runs. The manual for a particular engine tells the operator the rpm at which the engine should be running. The operator can easily overload an engine running too slow and damage it. Excessive centrifugal force can destroy an engine running too fast. Furthermore, if the engine also drives a generator, excessive speed can destroy the generator's windings. And, if the engine is driving a pump or a compressor, running too fast can damage the pump or compressor piston assemblies.

Tachometers

To summarize—

Engine instruments—Pyrometers

- measure the temperature of an engine's exhaust; exhaust temperature indicates whether engine is overloaded.
- measure the temperature of each engine cylinder; low cylinder temperature indicates an improperly firing cylinder; high cylinder temperature indicates an overloaded cylinder.

Oil-pressure gauges

- Oil-pressure gauges measure the pressure on the lube oil system.
- Low oil pressure can indicate (1) diluted oil, (2) low oil level, (3) blocked suction screen, and/or (4) engine bearing failure.
- High oil pressure can indicate (1) overly viscous oil or (2) improperly adjusted oil-pressure regulating valve.

Oil-temperature gauges

- Indicate lube-oil temperature.
- Lower-than-normal oil temperatures may allow water to condense or tolerances to vary.
- Higher-than-normal oil temperatures may cause oil to break down and fail.

Coolant-temperature gauges

- Indicate engine's operating temperature.
- Inlet coolant temperature should be no more than about 75°F (25°C) cooler than outlet temperature; a coolant that is too cold can crack the engine.

Air manifold-pressure gauge

- Indicates the pressure on the air-intake manifold
- Low manifold pressure indicates (1) dirty air filters or (2) malfunctioning supercharger (turbocharger).

Tachometers

- Measure engine rpm.
- An engine running too slow can be easily overloaded.
- An engine running too fast can destroy itself or equipment being powered by it.

Alarms and Shutdown Systems

Engine builders equip most engines with alarms and safety shutdown devices. Having an alarm sound or a shutdown occur does not necessarily mean that the engine operator is doing a bad job. Instead, alarms and shutdown devices prevent engine damage when an upset occurs. Most engines have two alarms: one for low oil pressure and the other for high coolant temperature. Also, most engines have three emergency *shutdowns*: one for low oil pressure, one for high coolant temperature, and a third for *overspeeding*.

Overspeeding alarms are not installed on engines because once an engine starts overspeeding, time is critical. By the time an operator could get to the engine upon hearing an alarm, it probably would have destroyed itself. A shutdown device is therefore essential to prevent overspeeding.

Should an engine overspeed, one kind of *overspeed trip device* immediately shuts off the fuel. Another kind of trip device shuts off both fuel and air. Operators should remember not to drop the load from an engine suddenly, because it may overspeed momentarily and trip the shutdown device.

Overspeed Trip Devices

The overspeed trip should be checked once a month to ensure that it is working properly. To check the trip, the engine is unloaded, or disconnected from whatever device it is driving, and its rpm is increased to the point at which it triggers the shutdown device. The required shutdown rpm should be indicated in the manufacturer's manual. If the trip fails to shut down the engine at a speed within 25 rpm of the manufacturer's specifications, the operator should not increase the engine's speed in an attempt to make the trip work. Instead, the mechanic should be informed of the problem.

Checking Overspeed Trips

Compounded Engines and Overspeeding

On rigs where the builders *compound* two engines, the operator should always run both engines in the same gear. If both engines do not run in the same gear, one engine can destroy the other by overspeeding it. To explain, let's say a rig has two compounded engines: one running in high gear, the other running in low. In this case, the engines run at different speeds to compensate for different gear ratios. The engine in high gear turns at a lower speed than the engine in low gear. Since the engines are compounded, the engine in high gear drives the engine in low gear, causing the engine in low gear to overspeed. Worse, even though the overspeed shutdown trips on the overspeeding engine, the engine cannot stop because the engine in high gear continues to drive it. The result is a destroyed engine.

Low-Oil Pressure Alarms

Most rig engines have low-oil pressure alarms. The alarm is usually a very loud horn. If the oil pressure drops below a preset amount, it trips a relay to sound the horn. On hearing an alarm, the operator should find the cause immediately. It may simply be that the operator needs to add oil. If not, however, the driller should be notified immediately. If the driller decides to remove the load from the engine with low oil pressure, the load should be quickly added to another engine. Otherwise, a critical rig component may not have enough power.

A low-oil pressure shutdown works in conjunction with the low-oil pressure alarm. The operator should set the shutdown to activate at a pressure lower than the alarm. This lower setting gives the operator time to remedy a malfunction. For example, if the engine's normal operating oil pressure is 60 psi (415 kPa), the alarm may sound when the oil pressure falls to 35 psi (240 kPa). Then, when the pressure falls to 25 psi (170 kPa), the shutdown activates. In this case, the engine continues to run at 25 to 35 psi (170 to 240 kPa) and the alarm sounds. But should the pressure fall below 25 psi (170 kPa), the shutdown device stops the engine and prevents damage to it.

The engine operator should turn off the low-oil pressure alarm and shutdown when starting an engine; however, these should be turned back on when the oil pressure reaches normal. This usually takes only about 30 seconds.

Most engines include a cutoff valve in a pressure line that runs to the alarm and shutdown solenoid. By closing the cutoff valve, the operator can test the alarm and shutdown devices. The closed cutoff valve disables the alarm and shutdown but does not affect engine oil pressure. To make the test, the operator unloads the engine and closes the cutoff valve. The alarm should sound first; then, as cutoff valve pressure continues to drop, the shutdown should stop the engine.

Low-Oil Pressure Shutdowns

Engine Start-up with Low-Oil Pressure Alarms and Shutdowns

Testing Low-Oil Pressure Alarms and Shutdowns

High-Coolant Temperature Alarms and Shutdowns

High-coolant temperature alarms and shutdowns work the same way low oil-pressure alarms and shutdowns do. If the coolant gets too hot, the alarm sounds. If the operator does not correct the problem, the coolant gets hotter, and the shutdown stops the engine.

In general, the operator can set the alarm to sound at about 195°F (90°C) and the shutdown to activate at about 210°F (100°C). Temperatures vary with the operating characteristics of the engine; the manufacturer or supplier should be consulted for the most accurate information on a particular engine.

Engine Shutdown Considerations

Under most circumstances, cutting off the flow of air to a diesel engine stops it. Also, cutting off the fuel supply can stop it. Sometimes, however, simply cutting off the fuel is not enough.

Blowouts and Natural Gas

One of the most dangerous situations on a rig is a *blowout*—the uncontrolled flow of well fluids such as natural gas into the atmosphere. During a blowout, natural gas can fill the air around the engines. Since all an engine needs to run is fuel, air, and ignition, shutting off the diesel fuel flowing to the engine in a gas-filled atmosphere may not shut down the engine. The engine takes in natural gas with the air, uses it as fuel, and continues to run. For this reason, some engines have an *air-shutoff valve* in the air-inlet manifold. The operator can use the air-shutoff valve to stop the engine manually or can set it to stop the engine automatically in case it overspeeds.

Turbochargers and Oil

Natural gas is not the only fuel that can enter with inlet air. On a turbocharged engine, an oil seal separates the lubricating oil from the intake air. If this seal leaks, the oil pump feeds oil into the intake air, and the resulting oil-and-air mixture fuels the engine. In this situation, shutting off the normal fuel supply does not stop the engine.

Using CO$_2$ to Stop an Engine

If an engine does not have an air-shutoff valve and the operator must cut off the air supply to stop the engine, the operator can use a carbon dioxide (CO$_2$) fire extinguisher. Spraying CO$_2$ into the air intake inlet displaces the oxygen in the air, and the engine stops firing. Under no circumstances should the operator use a dry chemical extinguisher. This will damage the engine.

To summarize—

Most engines have two alarms
- low oil pressure
- high coolant temperature

Most engines have three shutdown devices
- low oil pressure
- high coolant temperature
- overspeeding

Overspeed trip devices
- shut off fuel supply.
- shut off air supply.
- shut off both fuel and air supply.

Low oil-pressure alarms
- sound a horn if oil pressure falls below a preset level.

Low oil-pressure shutdowns
- shut down engine if oil pressure falls below a preset level.

High-coolant temperature alarms and shutdowns
- sound horn at a preset level.
- shut down engine at a preset level.

Engine Operation

Engine operators should follow proper procedures when starting an engine and putting it to work. Many rig owners and operators use the following steps.

1. All moving parts of the engine should be examined for proper adjustment, alignment, and lubrication. Parts to check include valves, cams, valve gear, fuel pumps, fuel-injection system, governor, lubricators, oil and water pumps, and the main machinery being driven by the engine.

2. The engine and machinery should be examined for loose nuts, broken bolts, loose connections, and leaks in jackets, joints, or valves. Nothing that must be tight should be loose and nothing that must be loose should be tight.

3. All pipes, valves, and ducts that carry fuel, oil, coolant, or air must be checked to make sure that they are not clogged, improperly adjusted, or dirty. If the engine has been idle for some time and is about to be put into service, the piping systems should be checked carefully for foreign matter. Many engine operators blow out the entire piping system with compressed air if the engine has been sitting idle for quite some time.

4. The engine should receive proper lubrication the moment it starts to turn. A complete checkup of the lubricating system is required to make sure oil is in every place where it is needed. Equally, all individually lubricated linkages and bearings must have an ample supply of clean oil, and all grease points must be properly lubricated.

Prestart Checks

5. The cooling system must be checked. If electric motors drive the coolant pumps, they should be started before starting the engine. It is essential that the coolant suction line be open so that coolant is in the engine. The operator can adjust the amount of coolant the pumps circulate while warming the engine.

6. The fuel system must be thoroughly checked. Pipes should be clean, pumps should be working, and the day tank should have fuel. It may be necessary to prime the fuel-injection pumps and release any trapped air from the discharge lines, valves, or nozzles. One or two strokes of the fuel-injection pump usually removes the trapped air. This should be done without forcing too much fuel into the engine's combustion chamber or cylinder. Too much fuel causes the engine to produce too much pressure on its first power stroke. Too much pressure, in turn, forces fuel oil into the crankcase, where it dilutes the lubricating oil. But while the operator should take care not to force fuel into the combustion chamber, it is important to put in enough fuel to fill each discharge line running from the injection nozzles.

7. If the engine has not been operated in some time, it should be turned over once or twice. To turn over the engine, the pressure-release devices on the cylinders should be opened. Releasing the pressure keeps the engine from starting. A bar inserted in the holes drilled in the rim of the flywheel enables the operator to turn it over by hand. Large engines may be turned over using the air motor or another air-operated device.

8. If the engine uses an air motor for starting, the pressure of the air in the storage tanks should be checked. If it is not at the required pressure, it should be pumped up to that level. Next, after closing the main control valve, the valve between the air tanks and the main control valve must be opened.

   If the engine uses a hydraulic starting motor, the pressure in the accumulator should be checked. If it is not at the correct prestart pressure, the accumulator must be charged.

If the engine uses an electric starter, the operator should make sure that the battery has a full charge.

9. The engine should not be loaded before starting. For example, if the engine drives a generator, the loading switch should be opened to disconnect the generator. If a friction clutch connects the engine to the load, the clutch should be put in neutral. If the engine drives a pump or a compressor, the bypass valve should be opened.

After completing the prestart check, the operator can start the engine, following these steps:

Start-Up

1. With air or hydraulic starters, the main starting valve is opened; on electric-start engines, the switch is turned on. More precise instructions may be found in the manufacturer's manual.

2. The operator must watch closely as the engine turns over. At the first indication of combustion, the starter motor should be cut off. An engine in good condition usually begins to fire between the second and fourth revolution of the crankshaft.

3. If the engine fails to start after four or five revolutions, the starter motor should be stopped. Rather than turning the engine over uselessly, the operator should try to find out why it is not starting.

Engine Warm-Up

After starting an engine, the operator must run it at the correct rpm in order to warm it up. The engine should not be run on idle to warm it up. An engine turning at idle speed does not burn enough fuel in the cylinders to heat the engine at an even rate. Uneven heating causes water to condense and collect in the crankcase. Water in the crankcase gets into the oil, mixes with sulfur and other contaminants, and forms a damaging sludge. Moreover, because of poor combustion when a cold engine is idling, the fuel does not burn completely. The unburned fuel gets past the piston rings and dilutes the lube oil. Eventually, the piston rings glaze (build up very hard deposits), which causes excessive oil consumption and blow-by.

The proper speed for an engine is between idling and full-rpm speeds. For example, if the engine normally runs at 2,000 rpm, a good warm-up speed is 1,000 to 1,300 rpm.

Warm-Up Speeds

The engine should reach operating temperature as quickly as possible, because cold oil cannot flow well enough to lubricate all the parts. Operators can do just as much harm, however, by running an engine too fast at start-up as by idling it for warm-up. Manufacturers make pistons from aluminum. As it heats up, aluminum expands almost three times faster than the iron in the engine block. Since the expansion rates are different, severe damage occurs to the pistons and block if an operator starts an engine and immediately runs it at high speed.

In a fast-running engine, the burning fuel contacts the top of the piston but does not contact a large area of the cylinder. The aluminum piston readily absorbs the heat of ignition; and as it does so, it expands. Unfortunately, the metal in the cylinder does not absorb heat at the same rate as the piston. As a result, the piston expands faster than the metal surrounding it and therefore enlarges before the cylinder does. Breaking through the oil film, the piston makes metal-to-metal contact with the cylinder wall. Metal-to-metal contact creates frictional drag, which generates such a high heat that it welds small bits of the piston to the cylinder wall. When the piston rings pass over these rough spots, the roughness damages the piston rings.

As mentioned earlier, the proper warm-up speed for a diesel is about half its rated running rpm. For example, if the engine's rated operating speed is 2,000 rpm, the operator should warm up the engine by running it from 1,000 to 1,300 rpm. The operator should not, however, run the engine at this speed for a long period; furthermore, no engine should be allowed to idle for extended periods at any speed. An engine should be started no more than about 30 minutes, but at least 10 minutes, before it is put to work.

During warm-up, the following observations should be made:

1. The operator should listen for normal combustion and the correct firing order. All cylinders must be checked for combustion and to determine whether the fuel-injection pumps are working properly.

2. The cooling system should be checked to make sure that the coolant pumps are working, the coolant level is correct, and coolant temperature is building up properly.

3. The operator should observe the oil pressure and oil temperature, feeling the cylinders to see whether any are warming up too fast. An insufficient amount of oil to any moving part spells trouble.

4. The exhaust stack should be examined to see if the color and sound of the exhaust are normal. The operator should make this check again after loading the engine.

A diesel engine should operate properly after it has run for about 5 minutes. If it is not running as it should, the operator should detect the malfunction within this warm-up period.

Putting an Engine to Work

Once the engine is properly warmed up, operators can put it to work. To properly load an engine, however, they must take into account what the engine is driving.

Engines Driving a Mechanical Assembly

If the engine drives a mechanical assembly, such as a transmission or a pump, operators can engage the engine by simply moving the engine clutch lever from the disengaged position to the engaged position. Once engaged, engine power sets the driven machinery in motion.

Engines Driving a Generator

If the engine drives a generator, the loading procedure is a little more complicated. The procedure varies, depending on whether the system produces alternating current or direct current. In either case, before starting the engine, operators should check the generator to make sure it is not connected to any electric load. If in doubt, they should consult with the rig electrician to determine the settings of all breaker switches and current relays before starting the engine or loading the generator.

Although many AC generators do not have rpm gauges (tachometers), it is critically important to monitor engine speed. Most AC generators in the U.S. produce 60-megahertz (MHz) current. A frequency meter measures and displays this current; it also regulates the engine's speed. Unfortunately, the frequency meter may indicate 60 MHz, not only at the proper rpm, but also at a lower rpm. If the meter does indicate 60 MHz at a low rpm, the engine may be overloaded because it has not reached its proper operating speed. Operators should, therefore, use a hand tachometer to determine the engine's actual rpm.

On a DC generator, operators should check engine speed with a tachometer each time they switch the generator to carry a load.

While the engine is running, the operator must stay on the lookout for leaks in the cooling system, in the injection valves, and in the air valves. Some leaks may not show up until a part has fully expanded, after the engine has been operating for a time with a normal load. No leaks of any kind must be permitted; if leaks do occur and cannot be corrected with the engine running, then the engine must be stopped while repairs are made.

In general, the operator should use the same procedures to check the engine while it is running under load as those used during warm-up. The operator should, however, observe the engine on a regular basis as it runs—for example, every half-hour. These regular observations should be made even if the engine has automatic alarms installed on it, and the readings should be entered in an engine report.

Checks to Make While Engine Runs

To summarize—

Prestart procedures

- Check all parts for proper adjustment, alignment, and lubrication.
- Check for loose parts.
- Check for clogged fuel, oil, and coolant lines, as well as clogged air ducts.
- Check for proper oil level.
- Check the cooling system.
- Check the fuel system.
- Turn the engine over manually if it has been out of service for a long period.
- Check the starting system.
- Ensure that the engine is not connected to a load.

Start-up steps

- Activate the starting valve or switch.
- When engine starts, shut off the starter motor.
- If the engine fails to start, stop turning it over and correct the problem.

Warm-up procedures

- Do not idle engine to warm it up.
- Run the engine at one-half its rated speed.
- Check combustion and firing order.
- Determine whether fuel pumps are working properly.
- Observe oil pressure and temperature.
- Check exhaust stack for normal exhaust.

Loading the engine

- Move clutch lever from disengaged to engaged position.
- If driving a generator, make sure the engine is running at the proper rpm.

Checks while engine runs

- Look for leaks and correct them.
- Observe the engine on a regular basis—every 30 minutes, for example.

Reports

Engine operators should keep accurate records of engine performance on a regular basis. By comparing engine logs, operators can spot trends that indicate abnormal performance and wear. Also, rig owners can compare the performance of different engines on different rigs and improve engine operations. What is more, engine manufacturers can get a good idea of how their engines perform, which often leads to performance improvements. Engine logs vary from rig to rig and company to company, but many rig owners include the following items on their reports.

1. *Time.* The time of day when writing down the engine readings is noted.

2. *Engine load.* In the case of an electric load, the volt and ampere readings are entered.

3. *Engine speed.* Engine speed is measured with a tachometer or an adding revolution counter. If using a revolution counter, the operator needs in addition a large electric clock with a hand indicating seconds. A large clock makes it easy to read the counter at exact time intervals.

4. *Fuel consumption.* If using a fuel meter to determine fuel consumption, the operator should take meter readings at exact time intervals.

5. *Exhaust.* Records should be kept on the—
   a. temperature of the exhaust from each cylinder;
   b. temperature in the exhaust line close to the exhaust manifold; and
   c. color of the exhaust, with a description such as clear, light haze, light gray, dark gray, or very dark gray.

6. *Lubricating oil.* Records should be kept on the—
   a. pressure of the oil as it leaves the oil pump discharge;
   b. temperature before it goes through the oil cooler;
   c. temperature after it goes through the oil cooler;
   d. pressure before it goes through the oil filters; and
   e. pressure after it goes through the oil filters.

7. *Coolant.* Readings should be taken of the—
   a. temperature of the coolant as it enters the cooling manifold; and
   b. temperature of the coolant as it leaves each cylinder or in the coolant outlet line.

8. *Scavenge air.* On two-stroke engines, the operator should note the scavenge air's—
   a. temperature after blowing; and
   b. pressure after blowing. Pressure readings are expressed in inches (millimetres) of mercury or psi (kPa).

9. *Supercharger condition.* Records should be kept on the—
   a. air temperature after the booster pump; and
   b. air pressure after the booster pump.

10. *Air temperature.* This reading should be taken at the intake in front of the air filter.

11. *Remarks.* Any notes concerning engine-related activities or any information that might be useful should be recorded. Some examples are: (1) put second engine on line; (2) stopped second engine; (3) found lubricating oil filter clogged as indicated by excessive drop, so switched to second filter; or (4) bypassed filter and changed filter element.

While the engines are running, operators should be alert for unusual sounds or knocks from the engine. They should also note whether the oil temperature is normal. High oil temperatures, for example, indicate that the engine bearings may be overheating. This suggests worn bearings that could lead to severe damage. Furthermore, operators should always make sure that the driller, or any other person, does not overload an engine. Finally, operators should check for overloaded cylinders. A cylinder with an exhaust temperature considerably higher or lower than normal is probably overloaded.

If the flow of oil or coolant stops for any reason, the entire engine, or perhaps only a cylinder or two, overheats. In either case, the engine should be stopped at once and allowed to cool gradually. Coolant should never be put into an overheated engine. The sudden temperature change can cause the pistons to seize (stop). Moreover, the cylinder heads, liners, or exhaust manifold may crack.

Engine Monitoring

Normally, an engine's exhaust should be perfectly clear. If, however, the driller overloads an engine, the exhaust may become visible as light gray smoke. Operators should never operate an engine for any length of time with a visible or smoky exhaust. Smoky exhaust may be the result of one or two cylinders behaving abnormally. Often, an increase in temperature indicates an abnormal cylinder. On the other hand, low cylinder temperatures mean that not enough fuel is getting to those cylinders and that the other cylinders are overloaded. If possible, the operator should stop the engine, find the cause, and fix it.

Exhaust Monitoring

Operators should maintain engines according to a schedule. The items on the schedule depend on the manufacturer's recommendations and on the equipment, instruments, and alarms the manufacturer provided with the engine. The type of service, the climate, and other circumstances also bear greatly on the maintenance schedule of an engine. When it comes to maintaining engines, operators must use the manufacturer's recommendations, the rig owner's recommendations, and any additional recommendations that may be in force for a particular rig at a particular time.

Maintenance Schedules

To summarize—

Items recorded on engine logs

- Time
- Engine load
- Engine speed
- Fuel consumption
- Exhaust temperature and color
- Lubricating oil pressure and temperature
- Coolant temperature
- Scavenge air temperature and pressure
- Supercharger condition
- Air temperature
- Remarks

Engine monitoring

- Be alert for unusual sounds.
- Check oil temperature.
- Check cylinder temperatures for overloaded cylinders.
- Coolant flow.

Exhaust monitoring

- Check for color of exhaust—it should be clear.
- Check for smoke.

Maintenance schedules

- Maintain engine according to schedule.
- Work out schedule according to manufacturer's and owner's recommendations.

ELECTRIC POWER
Introduction

As mentioned earlier, rigs transfer engine power to equipment in two ways: mechanical and electrical. On an electric-drive rig, the diesel engines drive generators, which produce electricity. Heavy-duty cables send the electricity to motors mounted on or near the equipment needing power. The electric motors power the equipment.

Generators and Alternators

Generators change mechanical power developed by the engines into electrical power. The first diesel-electric rigs used direct current (DC) generators. Drilling contractors still use DC rigs because many such rigs were built and are still in use. Today, however, most rig generators are alternating current (AC) generators, which generate AC electricity. AC generators are also called *alternators*.

Usually, AC or DC generators make electricity to power large DC motors, which are usually mounted on or very near the equipment they are powering. If the rig has AC generators, it also has equipment to convert *(rectify)* the AC to DC, since most motors operate on DC. Until recently, rig owners preferred DC motors to AC motors because DC motors develop the most torque at low speeds. Since rigs require a lot of power, often when the motors are fully loaded and turning slowing, DC motors won out over AC motors.

Lately, however, manufacturers have begun looking at AC motors to power equipment. AC motors are smaller and lighter than DC motors of comparable power. Further, when AC motors are equipped with modern electronic devices, the devices give drillers very precise control of the motor at all speeds. Moreover, AC motors can deliver higher torque or speed for a given horsepower. More work can be done using less energy, resulting in reduced fuel consumption and lower operating costs. Additionally, for a given horsepower rating, AC motors weigh less since they do not require the additional copper rotor windings and commutator bars required in DC motors. Since AC motors have fewer moving parts and no carbon brushes, maintenance is greatly reduced when compared to DC motors. In short, AC motors can perform better than DC motors under certain conditions, require less maintenance, and take less power to run.

Advantages of AC Generators

When it comes to generating electricity, AC generators are better than DC generators. Since most of today's rigs still have DC motors, it might at first glance seem inefficient to generate AC power and rectify it to DC, compared to generating straight DC power. However, generating AC power and rectifying it to DC has advantages. For one thing, rig equipment needs a lot of horsepower *(kilowatts)* to operate. Because of the way AC generators work, manufacturers can build AC generators bigger, cheaper, and more powerful than DC generators. As a result, rig owners can use more powerful diesel engines to drive the big AC generators, which means that they can use fewer engines and generators to power the rig. Power costs are thus reduced.

For another thing, many small motors on the rig, plus the rig's lighting, require AC power. If the rig generates DC power, then the rig owner must use auxiliary AC generators for the lights and motors. With AC generators, the rig does not require auxiliary generators.

Because they do not use commutators, AC generators require less maintenance than DC generators. *Commutators* are rotating segmented rings that tend to wear out the brushes that touch them. Since manufacturers use a continuous ring in AC generators, the brushes do not wear as fast.

DC Generators

Since many DC-to-DC rigs are still in operation, they are worth studying. Figure 71 diagrams the operation of a simple DC generator. A rectangular loop of stiff copper wire rotates between the north and south poles of an electromagnet that creates lines of force called *magnetic flux.* These lines of force flow from the magnet's north pole on the right to the south pole on the left. Parallel dashed arrows show the magnetic flux flowing between the two poles.

As the loop rotates through the magnetic flux, it generates electric current. Each end of the loop connects to a commutator. The simple commutator in the schematic diagram is composed of two segments of copper; in reality, commutators are made up of dozens of segments. Each half of the commutator segment in figure 71 rotates against a *brush*—a small, flexible bar made of a material that conducts electricity. The brushes conduct electricity to a circuit outside the generator. In figure 71, the current flows to a meter. The meter's pointer deflects to the right to show current flow.

In figure 71*A*, the black side of the loop is moving up, while the white side is moving down. Current flows clockwise through the loop. It goes to the commutator segment on the right, through the brush, and to the meter. The pointer on the meter shows that current is flowing.

In figure 71*B*, the loop has rotated so that it is perpendicular to the magnetic flux lines. No current flows at this particular moment. The commutator segments are not touching the brushes, so the meter shows that no current is flowing.

In figure 71*C*, the loop has turned 180 degrees. The current, however, still flows clockwise through the loop, so the meter's pointer still deflects to the right as it did in the top drawing. This simple generator therefore generates direct current.

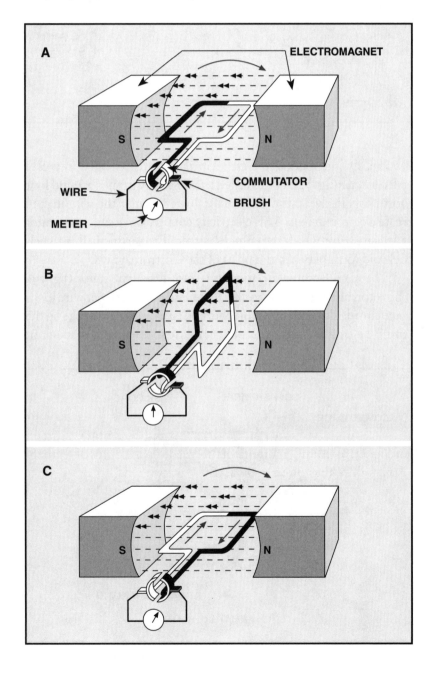

Figure 71. Simple DC generator

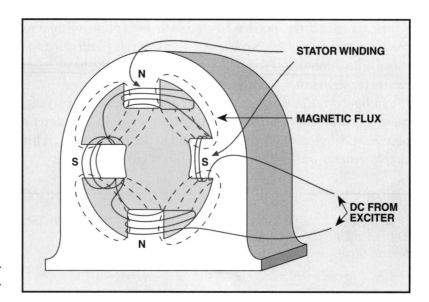

Figure 72. Generator stator

In a DC generator, the electromagnets are arranged in a case called a *stator* (fig. 72). Wires called *windings* are wrapped around four short posts in the stator. Electricity flows through the windings to create electromagnets. This electricity comes from a small generator called an *exciter*, which the engines power. The magnetic flux travels back and forth between the poles of the electromagnets.

The manufacturer also mounts a rotor inside the stator (fig. 73). The rotor, also called an *armature*, contains several copper wire loops placed inside slots cut into a core. Several laminated disks make up the core. The ends of each loop connect to the segmented commutator.

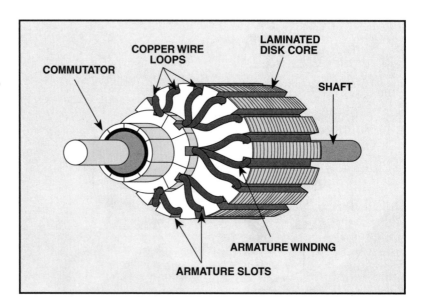

Figure 73. Generator rotor (armature)

AC Generators

Figure 74 diagrams the operation of a simple AC generator (alternator). Like the DC generator, it has an electromagnet that generates magnetic flux from a north to a south pole as indicated by the parallel arrows. It has a single copper wire loop rotating clockwise through the magnetic flux. Each end of the loop connects to a slip ring, which is in contact with brushes. Unlike the segmented commutator, the *slip ring* is a continuous ring. A wire from each brush conducts current to a meter.

In figure 74A, the black side of the loop is rotating upward through the flux. At the same time, the white side is rotating downward. Current flows through the gray slip ring, makes contact with the brush, and flows into the wire circuit and meter. At this moment, current flows clockwise through the circuit as indicated by the arrow and the meter pointer deflecting to the right.

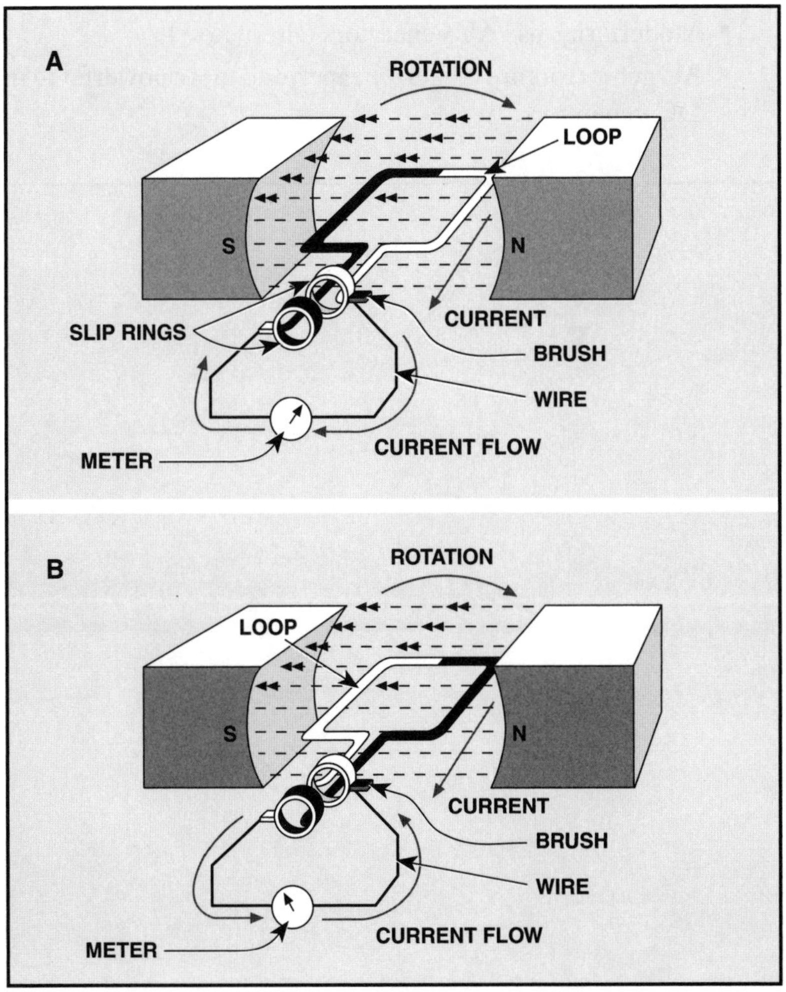

Figure 74. Schematic of a simple AC generator (alternator)

In figure 74B, the white side of the loop is rotating upward through the flux, while the black side is going down. Current now flows through the black slip ring, which touches the brush, and flows into the wire and meter. At this moment, current flows counterclockwise through the wire and meter as indicated by the arrow and the meter pointer deflecting to the left.

Because the loop alternates in contacting two separate slip rings, the current flow changes from clockwise to counterclockwise on each rotation of the loop.

To summarize—

Generators and Alternators

- Generators change mechanical power to electrical power.
- Modern rigs use AC generators (alternators).
- AC generators are bigger, cheaper, and more powerful than DC generators.

DC Electric Drive

The first diesel-electric rigs routinely started drilling holes in the 1960s. They rapidly began replacing mechanical rigs, especially offshore. On these early electric-drive rigs, diesel engines drove DC generators coupled to the engines. The direct current went through heavy-duty controls, switches, and electric cables to DC motors. The rig builder mounted the DC motors on or near the equipment requiring power. Today, many rigs still use DC-to-DC electric drive.

Converting mechanical power produced by an engine into electrical power and then back into mechanical power may seem like a long way around, but it has advantages. Unlike a mechanical rig, where machinery transfers engine power, on an electric rig the flow of power from the engines to the driven equipment is smooth. The driven machinery delivers no shock back to the engine. A DC motor produces its greatest torque when it stalls, and the engines continue to put out full torque even if the motor stalls. As a result, a lot of torque is available to turn the equipment at very low speed and under heavy load.

When an operation needs a lot of power, the driller can switch (assign) all the electric power the generators put out to one piece of equipment. For example, to raise the mast, the driller runs the engines at full speed, puts a full field on the motor's electromagnets, and varies the electromagnetic field in the generators to control (accelerate and decelerate) the equipment the motor drives.

Advantages of Electric Drive

177

Exciters

The rig builder mounts a small DC generator, called an exciter, on or near the main generator. Whether the exciter is run by an engine or an outside source, it puts out direct current. This current goes to the stator in the main generator to produce the electromagnetic field. The driller then varies the output of the exciter from the console on the rig floor. By varying the exciter's output, the driller controls the strength of the magnetic field. The stronger the magnetic field of the generator, the more current the generator puts out to drive the load.

Running Engines at Constant Speed

On one type of DC-DC electric rig, the engines run at a constant speed. This constant engine speed runs the exciters at a constant speed. The driller varies the current from the exciter by using a rheostat. (A *rheostat* is a variable resistor that controls the amount of current flowing in a circuit.) The driller adjusts the rheostat to control resistance in the flow of electricity going to the generator's electromagnetic field. A lower resistance allows more current to flow. The more current the exciter delivers, the stronger the field in the generator, and the more current the generator puts out to drive the load.

Running Engines at Varying Speeds

On another type of DC-DC electric rig, the driller varies the speed of the engines. Varying engine speed varies the speed of the exciter, which, in turn, varies the strength of the generator field. In this setup, the same engine that drives the generator also drives the exciter.

A schematic diagram of a typical DC-DC control hookup (fig. 75) shows two ways to generate electricity. At the top of the drawing Engine 1 drives generators 1 and 2, while Engine 2 drives generators 3 and 4. In this case, the driller holds the engine speeds constant. The scalloped lines just below each generator represent the windings in the exciter. The "half-clock" symbols next to them represent the rheostats that vary the flow of current from the exciters to the generators. The small white rectangles below the rheostats represent the air actuators or electric actuators that operate the rheostats.

DC-DC Schematic

Figure 75. Control hook-up of DC electric drive, typical except for air throttle rather than more common electric throttle

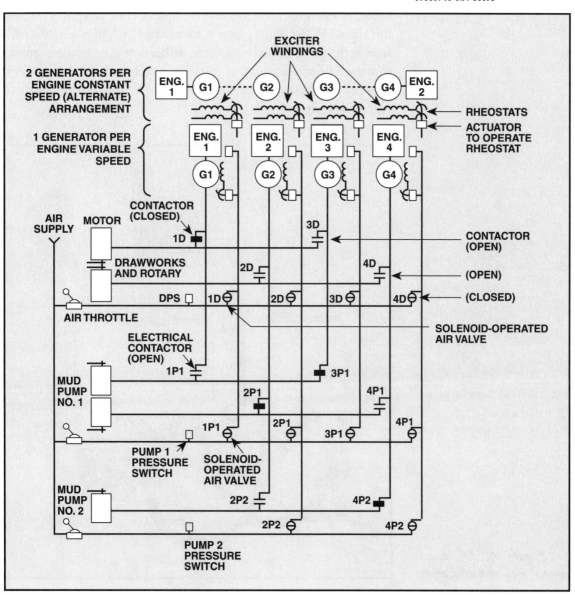

179

The larger white rectangles represent engines in an alternative system that has an engine for each generator. In this system, the driller varies the speed of the engine to vary the exciter's magnetic field. The system that regulates the engine throttle also controls the rheostat on the exciter.

The drawworks motors diagrammed in figure 75 also drive the rotary table. Another arrangement uses a separate motor to drive it. In still other cases, the rig may use a top drive, which does not require a rotary.

The generators shown in figure 75 can deliver up to 600 hp (420 kW). The five motors can put out 625 hp (438 kW) continuously, as when running the mud pumps, or they can put out 1,000 hp (1,400 kW) for short periods, for example, when raising the drill stem at the hoist. In such a case, the driller would tie two generators into each drawworks motor at 1D, 2D, 3D, and 4D in figure 75. Typically, rig owners use 625-hp (438-kW) DC motors to drive equipment (fig. 76).

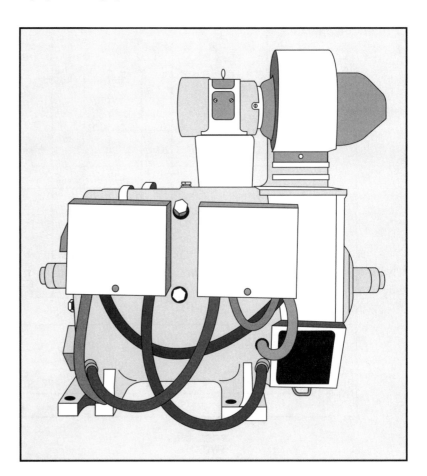

Figure 76. DC electric motor used on a drilling rig

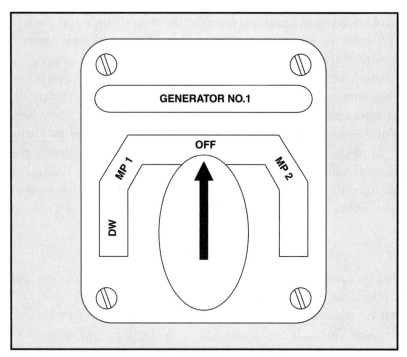

Figure 77. Assignment switch for DC electric drives

The driller uses a switch called an assignment switch on the driller's console to activate each generator (fig. 77). To "make an assignment," the driller switches on the generator. The assignment switch sets up the speed control and power circuits and puts a full field on all the motors to which the driller assigns power.

Assigning Power

Figure 75 provides an example of assigning power. Note the closed electrical *contactors* indicated by small black rectangles. With contactor 1D closed, the driller assigns power from generator 1 to drive the rotary table. With contactors 2P1 and 3P1 closed, generators 2 and 3 drive mud pump no. 1. With contactor 4P2 closed, generator 4 drives mud pump no. 2.

Using the Throttles

Still referring to figure 75, note the three air throttles on the left. To start the rotary table, the driller advances the air throttle controlling the drawworks and rotary table. The initial air pressure in the throttle line causes the drawworks pressure switch to (1) close the contactor on the line from generator 1, (2) open the solenoid-operated air valve in the air line, and (3) close the other air valves. Further advancement of the air throttle increases the magnetic field in the generator in direct relation to the movement of the throttle handle, and the rotary begins to turn at the speed set on the air throttle. Similarly, the driller uses an air throttle to start generators 2 and 3 and adjust the speed of pumps 1 and 2.

Driller's Control Panel

Except for the assignment switches and the speed controllers, the driller's control panel contains air controls like those on a mechanical rig (fig. 78). Other devices on the control panel include an *ammeter* for each motor. An ammeter measures the amount of electrical current produced by the motor in units called *amperes* (amps). Since the amperage put out by the motor translates into torque, the driller can use the ammeter to measure the torque each motor develops.

A *voltmeter* on the panel indicates engine speed by measuring electrical voltage. *Voltage* is somewhat like the pressure on fluid flowing in a pipe. In the case of electricity, voltage is electrical force. The higher the voltage in a circuit, the higher its electrical force. And, the higher the electrical force, the faster the engine runs.

The panel also includes a reversing switch for the drawworks and rotary, an emergency stop button, and indicating lights. As mentioned earlier, the panel also has an assignment switch for each generator.

ASSIGNMENT SWITCHES

Figure 78. Control panel for DC electric drives

Main Control Cabinet

A control cabinet houses most of the electric and pneumatic controls for a DC-DC drive. Cables come out of the control cabinet and carry the circuits necessary to control the rig. All the electrical connections are of the plug-in type. When rigging up, personnel lock the connections in place to prevent them from disconnecting (fig. 79).

The control cabinet also houses all the safety devices needed to shut down a circuit in case of a short, overload, or damaged cable. Interlocks set up a sequence of events that prevents damage to, or operation of, equipment should someone engage the wrong operating lever.

A ground relay shuts off power if a cable is damaged and sends electricity through any structure that is not part of the electrical system. This system prevents electrical shock to operating personnel during normal operations. It is, however, vitally important to follow proper lock-out and tag-out procedures when maintaining or repairing any of the equipment.

Figure 79. Plug-in connector

Dirt and dust are the enemy of all electrical equipment, so the equipment must be kept as clean and as dust free as possible. Dirt collects moisture, which causes short circuits and binding of moving parts. Oil and grease are good dirt collectors, so they should be kept away from electrical equipment. Following the manufacturer's recommendations when greasing generator or motor bearings is very important for proper maintenance.

A lintless cloth is the best medium for cleaning large equipment. Dirt may also be blown away with dry compressed air. If it becomes necessary to use a solvent, the operator should make sure that the solvent is approved for electrical equipment; some solvents dissolve insulation. Water should never be used to clean electrical equipment. Materials dissolved in even the cleanest water conduct electricity and cause short circuits. Everyone who works around electrical equipment should keep in mind that water and electricity do not mix.

A small paint brush serves to brush dirt from small equipment. Once the dirt is brushed off the equipment, a vacuum cleaner may be used to remove it. Dirt should not be allowed to fall into contacts and armatures.

No oil or grease should be used on a commutator, whether it is part of a motor or a generator. A commutator operating normally has a film on it, but this film never needs oiling. Grooved or streaked commutators should be repaired or replaced.

Burned or broken brushes in a generator or motor mean that the commutator is out of round or has high or low commutator bars. Crew members must pull such a unit and send it to a repair shop. Most likely the shaft is bent or the motor has been overheated.

Maintenance

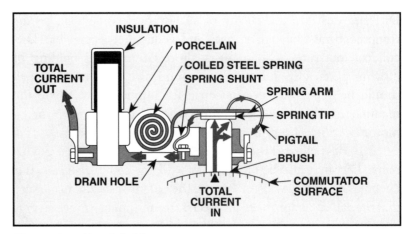

Figure 80.   Brush holder

Motor and generator brushes should be regularly inspected. Carbon dust that collects in the brushes should be blown away; this will keep the brush from binding in its holder. Brush holders should be kept in good repair. In one type of brush holder (fig. 80), current enters a brush from the commutator. Most of the current then flows through the *pigtail*. When operating normally, the pigtail makes a solid connection between the brush and holder. The *spring shunt*, which is a twisted wire similar to the pigtail, acts as a backup if anything goes wrong with the pigtail. The shunt diverts heavy current around the coiled steel spring to avoid damage by overheating. If overheating occurs, the brush loses contact pressure against the commutator, which causes the generator or motor to operate improperly.

When the rig is shut down, all connections must be checked for tightness. If they are loose, the terminals will be hot or discolored. Frayed shunts, burned arc chutes, broken springs, and broken terminal strips should be replaced. Resistor tubes should be checked for cracks, and any rough or worn contact surfaces on contactors, relays, and interlocks should be replaced.

SCR Systems

Diesel-electric rigs using diesel engines to power DC generators, which, in turn, powered DC motors, were the first on the scene. In the late 1960s, when transistors and other solid-state electronic devices came into their own, rig designers and manufacturers began to develop other ways to transmit and control electric power. They began using diesel engines to turn AC generators, converting or rectifying the AC to DC, and using the DC to power motors on the equipment. Solid-state devices called *silicon-controlled rectifiers* (SCRs) rectify AC to DC. SCRs are also called *Thyristors*.

SCRs or Thyristors

An SCR or Thyristor is related to a solid-state device called a *diode* or a *rectifier*. It allows current to flow in only one direction. Like a check valve in a water pipe, the diode lets AC current flow in one direction only, thus turning it into DC.

An SCR differs from a simple diode, however. Unlike a diode, an SCR will not conduct electricity in either direction until it receives control voltage. *Control voltage* is electricity applied to the SCR from a separate source, usually from a small generator run by a rig engine. This control voltage is often called *gate voltage*, because, like a gate that can open or close, the SCR can be made to conduct or not conduct electricity, depending on whether control voltage is applied.

When the SCR gate is energized by the relatively small control or gate voltage, the SCR conducts electricity in one direction. When the gate is not energized, the current is blocked in both directions. The SCR and its controlling voltage can operate very rapidly—up to thousands of times per second. Thus, the SCR gate opens and closes rapidly to allow and stop current flow rapidly. This rapid opening and closing allows the driller to control the power going to the DC motor.

Controlling Power

AC power, or voltage, is pulsating in nature. A technician can measure AC voltage by hooking it up to an *oscilloscope*, which is somewhat like a small TV set. An oscilloscope projects a picture of AC voltage in the shape of a wavy line that looks similar to an ocean wave. Called a *sine wave*, the line pulses above and below a base line (fig. 81).

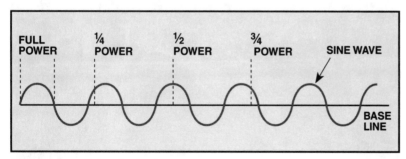

Figure 81. DC voltage shown as a sine wave

If the driller makes the SCR conduct electricity all the time, the SCR relays full current in only one direction, thus converting AC to DC at maximum power capacity. The sine wave diagrammed in figure 81 indicates maximum power if DC flows the entire time the wave is above the base line. If, however, the driller makes the SCR conduct during only part of each sine wave, the SCR converts and transmits only part of the current at reduced power. By using a special controller, the driller can govern the point on the sine wave when the SCR starts to conduct, and thus vary from zero to maximum the amount of rectified power relayed to the motor.

To satisfactorily control the gate voltage, the manufacturer has to combine several input signals. For example, the driller has to be able to control a motor's power output, so the manufacturer includes a throttle signal. Other signals fed to the gate-control system include (1) current-limit signals to limit the maximum amount of current conducted; (2) a rate-limit signal to control the rate of current turn-on, which minimizes electrical surges; (3) a paralleling signal to allow the rig operator to run more than one SCR in parallel (paralleling SCRs may be required to power a large motor); and (4) motor speed-limit signals to prevent the motor from being run too fast.

AC Bus and Control Units

In the SCR (AC-DC) system, standard three-phase alternators generate AC power (fig. 82). The size and type of rig determine the size and voltage of the alternators. Heavy-duty cables feed power from the alternators to a common set of conductors.

Figure 82. Three-phase alternators connected to rig engines

This set of conductors is called the *AC bus* or the *common bus* (fig. 83). The common bus is made up of large copper cables, usually called *bus bars*. Power from the bus then flows to special units or cabinets, where the SCRs rectify the AC and where other controls are located.

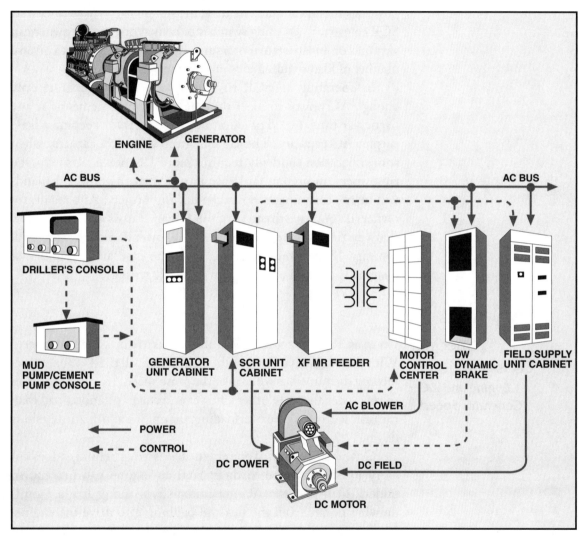

Figure 83. SCR (AC to DC) rig system

Meeting Power Needs

Each motor on an AC-DC electric rig usually drives a load that is different from the load on other motors in the system. In other words, each motor's load is unique to that motor. The manufacturer provides a device called an SCR converter that a technician can set to provide the proper voltage and power for each of the different loads. But each motor must have its own SCR converter, and if the driller needs to power six motors at the same time, the system has to have at least six SCR converters. By using switches or contactors, the rig owner can vary the combinations of motors and SCR converters so that a diverse number of load requirements may be met with ease.

In operating an SCR rig, the system has to generate only enough AC power to meet the total DC-motor demands at any particular time. Good practice, however, calls for keeping a little surplus AC capacity. The rig may need this extra capacity when some operation suddenly requires more DC power. Usually, the rig owner can predict fairly accurately what the power demands will be for each operation in the drilling project. As a result, rig owners do not have to run all of the AC generators when they know that a particular operation requires less power than the system's full capacity. This economy, along with the ease and flexibility of control, is the real advantage of the SCR system.

Special Considerations
Engine and AC Generator Speed

On an SCR rig, it is essential for all engines to run at the same speed. With the engines running at the same speed, all the AC generators feeding the same bus will run at the same speed. If one engine tries to run faster than the others, however, it takes on more load than the rest. With extreme overloading, the engine automatically shuts down.

On the other hand, if one engine tends to lag behind, it does not carry its share of the load. In fact, an engine can become so underloaded that the AC generator starts acting like a motor, drawing power from the bus and beginning to drive the engine. Running a diesel engine under a no-load condition with a motor driving it can destroy it. To prevent such a condition, manufacturers install a *reverse current relay*. This relay senses current flowing in the wrong direction—that is, current flowing from the generator back to the engine, instead of from the generator to the bus. When the relay senses reverse flow, it disconnects the generator from the bus. Disconnecting the generator prevents it from acting like a motor and thus saves the underloaded engine from damage.

Because it is so important that all engines and AC generators carry equal shares of the load, the manufacturer installs very sensitive and accurate engine governors. Rig technicians should keep a close eye on the governors to ensure that they are working correctly to keep the engine speeds constant. The technician can manually adjust the governors if necessary.

When the total load on the engines is relatively constant, the technician can easily adjust each engine governor so that each generator set carries its share of the total load. When a rapid and great change in load demand occurs, however, the system may be fully loaded at one instant and nearly unloaded at the next. To meet such emergencies, SCR rigs use integrated, electronically controlled governors on each engine. Electronic governors sense the output current of each generator and automatically and rapidly adjust engine speeds to ensure that each engine carries an equal share of the load.

Electronic Governors

Rig operators need to take care when adding or removing AC generators from the line. As emphasized above, all generators connected to the common bus must run at identical or *synchronous speed*. If they do not run at synchronous speed, any "out-of-sync" generators can be severely damaged or destroyed. When a generator is said to be running at synchronous speed, this means that it may be running faster or slower than other generators, but it is running in cycle with them. That is, its electrical output is in phase with the other generators. The sine waves produced by generators running "in sync" all match.

In any case, when rig operators put an additional generator on line, they must either bring it up to the exact speed of the other generators or bring it to a speed that synchronizes it with the other generators, before they close the switch connecting it to the bus.

The following steps should be taken to bring an AC generator on-line.

1. The engines are started and allowed to warm up.
2. The control on the governors is adjusted until each engine's speed is the same as the others.
3. The voltage control for the generator being brought on line is adjusted to match the voltage of the other generators.
4. Using the synchronizing panel on the AC switchboard, the governor control is adjusted so that the engines are running the generators at the exact synchronous speed.

Adding or Removing Generators from the Line

The moment when the generators have achieved exact synchronous speed is determined using the *synchroscope*, an instrument on the synchronizing panel.

5. When synchronous speed is achieved, the generator's circuit breaker is turned on (thrown) to connect the generator to the bus. The operator should never throw the breaker when the generator phases are out of synchronization; this will cause a severe jolt to the out-of-sync generator that is likely to damage it.

6. When first put on line, the generator does not carry a load. To load it, the operator speeds up the engine by adjusting the engine governor. As the engine speeds up, the generator picks up more and more of the load. Finally, all the ammeter readings on the generators are compared to ensure that the generators are all carrying the same load. Governors may be adjusted, if necessary, until all generators are putting out the same amperage. (As mentioned earlier, an ammeter measures and indicates the amount of amperes—amps—a generator puts out. An ampere is a measure of the amount of electrical current flowing in a conductor. The more current a generator puts out, the higher the amperage.)

To remove an AC generator from the line, the procedure is reversed.

1. The governor control for the particular engine is retarded until no current is being produced by the generator. The ammeter on this generator should be consulted to determine when current production stops.

1. The circuit breaker on the generator is turned off (tripped) to remove the generator from the line.

3. The engine is shut down in the usual manner.

An improperly set governor or an automatic engine shutdown can create a problem. As mentioned earlier, improper engine speed or a total shutdown causes the generator to act like a motor. Like a motor, the generator then starts drawing current from the bus, rather than supplying it. In such a case, the reverse current relay trips the generator circuit breaker and automatically removes the generator from the line. If the rig's total power demand at that moment is such that the remaining generators can carry the load, nothing further happens. If, however, the remaining generators do not have the capacity to meet the demand, they automatically shut down. As a result, no generator feeds power to the bus.

Lack of rig power is, at best, troublesome, but it can also be disastrous. To prevent a complete rig shutdown, rig owners can incorporate overload protection into their rig's SCR system. An overload protection system reduces all the power to the loads by the same percentage. Reducing power by the same percentage keeps the total power demand from exceeding the generating capacity.

One overload protection system works by means of a sensing module which receives input on the capacity of the generating units on line. It also receives information on the total load being drawn. The module compares the load being imposed with the capability for serving this load. If the load gets too close to the capability, the module develops an electrical signal which it sends to all SCR systems in operation. This signal causes each system to reduce its output, thereby reducing the total power requirement. The reduction is done on a percentage basis, so no loss of any load occurs; rather, the power delivered to each is reduced.

Malfunctions

Auxiliary Power Supply   Mechanical and DC-DC rigs require auxiliary equipment to generate AC power. AC provides rig lighting and runs several small motors on the rig. (Recall that SCR rigs, since they generate AC, do not require auxiliary AC power generation equipment.)

The total AC electrical power requirements for the auxiliary equipment on a small-to-medium-size land rig may not be more than 100 kW. Rigs usually have enough AC generating capacity to provide twice the amount of power required. Having twice the power required means that the rig has 100 percent standby generating capacity. That is, if half the generators failed, the rig would still have enough power to generate full capacity. Table 1 shows some of the main items requiring AC power on a land rig.

Table 1
AC Power Required By Rig Equipment

Item	Power (kW)
Electric lighting	12
Shale shaker (two 3-hp or 2.1-kW motors)	5
Mud-tank agitators (four 10-hp or 7-kW motors)	30
Desander, centrifugal pump (35 hp or 24.5 kW)	25
Degasser and centrifuge unit (12 hp or 8.4 kW)	10
Air compressors (25-hp or 17.5-kW motor)	20
Bunkhouse cooling and heating	5
BOP accumulator unit	10
Mud accumulator unit	10
Miscellaneous	5
Total	132

To summarize—

DC-DC Electric Drive

- Diesel engines drive DC generators.
- Cables and switch gear carry DC power to DC motors on equipment.
- Driller can run engines at constant speed or at different speeds.

Rules for Maintenance

- Keep equipment clean.
- Blow dirt away on large equipment.
- Brush away and vacuum dirt on small equipment.
- Do not use water to clean.
- Check connections for tightness when rig is shut down.
- Do not use grease or oil.

SCR Systems

- Diesel engines drive AC generators (alternators).
- Silicon controlled rectifiers (SCRs) or Thyristors convert (rectify) AC to DC.
- DC motors are mounted on equipment to be powered.
- All engines should be run at the same speed.
- All AC generators should be run at the same or synchronous speeds.

Glossary

AC bus *n*: in a diesel-electric power system, a common set of conductors made up of large, heavy-duty copper cables that carry alternating current generated by the system's alternators (AC generators).

acid corrosiveness *n*: a characteristic of diesel fuel that indicates the likelihood of a diesel fuel's causing corrosion as the engine burns fuel. In general, a fuel with a high acid content will be more corrosive than a fuel with low acid content.

aftercooler *n*: on a supercharged engine, a device, cooled by either air or by engine coolant, that reduces the temperature of the engine's exhaust. It is necessary to cool the exhaust's temperature because the exhaust drives the supercharger, which forces air into the engine's intake manifold. The temperature of the supercharged air must be at an acceptable level; otherwise, the engine will run too hot. See *supercharger*.

air intake manifold *n*: on a diesel engine, an arrangement of pipes and passageways through which air is conducted to the engine's combustion chambers.

air knocking *n*: on a diesel engine, a phenomenon that occurs when trapped air in the fuel injection system enters the engine's cylinder with the fuel. The fuel-air mixture ignites but, because of the extra air in the fuel, the engine cylinder misfires and knocks or hammers. The problems should be corrected promptly to prevent damage to the engine.

air-motor starter *n*: on an engine, a device powered by compressed air that starts the engine. The compressed air, when allowed to enter the starter motor by means of a valve, causes a gear on the starter to engage a gear attached to the outer edge of the engine's flywheel. The rotating starter gear moves the flywheel gear, which causes the engine's pistons to move. If fuel, air, and, on spark-ignition engines, an electric spark are present in the engine, the engine will start after a few rotations. As soon as the engine starts, the starter gear disengages from the flywheel gear. Air-motor starters are installed on large industrial engines like those used on a drilling rig.

air shutoff valve *n*: on a diesel engine, a special valve that, when activated, prevents air from entering the engine's combustion chambers, thereby stopping the engine. Air shut-off valves are a safety feature that may be needed when a well blows out. If natural gas is present in the blowout's fluids, a diesel engine can take in the gas and continue to run even when its normal fuel source is cut off.

alternating current (AC) *n*: current in which the charge-flow periodically reverses and whose average value is zero. Compare *direct current*.

alternator *n*: an electric generator that produces alternating current.

American Petroleum Institute *n*: founded in 1920, this national oil trade organization is the leading standardizing organization for oilfield drilling and producing equipment. It maintains departments of transportation, refining, marketing, and production in Washington, DC. It offers publications regarding standards, recommended practices, and bulletins. Address: 1220 L Street NW; Washington, DC 20005; (202) 682-8000.

ammeter *n*: an instrument for measuring electric current in amperes.

ampere (A) *n*: the fundamental unit of electrical current; 1 ampere = 6.28×10^{18} electrons passing through the circuit per second. One ampere delivers 1 coulomb in 1 second.

annular blowout preventer *n*: a large valve, usually installed above the ram preventers, that forms a seal in the annular space between the pipe and the wellbore or, if no pipe is present, in the wellbore itself. Compare *ram blowout preventer*.

antifreeze *n*: a chemical added to liquid that lowers its freezing point. Often used to prevent water in an engine's cooling system from freezing.

API *abbr*: American Petroleum Institute.

API gravity *n*: the measure of the density or gravity of liquid petroleum products in the United States; derived from relative density in accordance with the following equation:

$$\text{API gravity at } 60°F = [141.5 \div \text{relative density } 60/60°F] - 131.5$$

API gravity is expressed in degrees, 10° API being equivalent to 1.0, the specific gravity of water.

armature *n*: a part made of coils of wire placed around a metal core, in which electric current is induced in a generator, or in which input current interacts with a magnetic field to produce torque in a motor.

aspiration *n*: in internal combustion engines, the method which the engine uses to take air into its cylinders. Engines can be naturally aspirated, which means that the pistons draw in air at atmospheric pressure as they move down the cylinder; or they can be blown or supercharged, in which an engine-driven compressor of some type raises the pressure of the air above that of the atmosphere and forces it into the cylinder.

atmospheric pressure *n*: the pressure exerted by the weight of the atmosphere. At sea level, the pressure is approximately 14.7 pounds per square inch (101.325 kilopascals), often referred to as 1 atmosphere. Also called barometric pressure.

atomize *v*: to spray a liquid through a restricted opening, causing it to break into tiny droplets and mix thoroughly with the surrounding air.

B **back-pressure** *n*: 1. the pressure maintained on equipment or systems through which a fluid flows. 2. in reference to engines, a term used to describe the resistance to the flow of exhaust gas through the exhaust pipe. 3. the operating pressure level measured downstream from a measuring device.

baffle plate *n*: 1. a partial restriction, generally a plate, placed to change the direction, guide the flow, or promote mixing within a tank or vessel. 2. a device that is seated on the bit pin, in a tool joint, or in a drill pipe float, to centralize the lower end of a go-devil while permitting the bypass of drilling fluid.

bag filter *n*: on an engine, a bag-shaped piece made of cotton or fiber cloth that fits into a special holder in the fuel system piping. Fuel is circulated through the bag, which removes foreign matter from the fuel.

ball bearing *n*: a bearing in which a finely machined shaft (a journal) turns on freely rotating hardened-steel spheres that roll easily within a groove or track (a race) and thus convert sliding friction into rolling friction. See *ball race*.

ball race *n*: a track, channel, or groove in which ball bearings turn.

bar *v*: to move or turn (as a flywheel) with a bar used as a lever.

battery *n*: 1. an installation of identical or nearly identical pieces of equipment (such as a tank battery or a battery of meters). 2. an electricity storage device.

BDC *abbr*: bottom dead center.

Bendix *n*: the brand name for a type of friction clutch in an electric starter for small engines. When electric current is applied to the starter, the friction clutch (the Bendix) moves forward to engage a pinion gear on the starter with a ring gear on the engine flywheel. As the starter's pinion rotates, it rotates the ring gear, which moves the flywheel to turn the engine over.

blow-by *n*: the percentage of gases that escape past the piston rings from the combustion chamber into the crankcase of an engine.

blowout *n*: an uncontrolled flow of gas, oil, or other well fluids into the atmosphere. A blowout, or gusher, occurs when formation pressure exceeds the pressure applied to it by the column of drilling fluid. A kick warns of an impending blowout. See *kick*.

blowout preventer *n*: one of several valves installed at the wellhead to prevent the escape of pressure either in the annular space between the casing and the drill pipe or in open hole (i.e., hole with no drill pipe) during drilling or completion operations. Blowout preventers on land rigs are located beneath the rig at the land's surface; on jackup or platform rigs, at the water's surface; and on floating offshore rigs, on the seafloor. See *annular blowout preventer, ram blowout preventer*.

booster pump *n*: on a diesel engine, a small manually or electrically operated pump that an engine operator can use to prime the engine's fuel system for starting the engine. When activated, the pump moves fuel from a tank, through the engine's fuel lines, and to the engine's injectors, ensuring that fuel is available for starting the engine.

bottom dead center (BDC) *n*: the positioning of the piston at the lowest point possible in the cylinder of an engine; often marked on the engine flywheel.

breather *n*: a small vent in an otherwise airtight enclosure for maintaining equality of pressure inside and outside.

brush *n*: a carbon block used to make an electrical connection between the rotor of a generator or motor and a circuit.

bus *n*: an assembly of electrical conductors for collecting current from several sources and distributing it to feeder lines so that it will be available where needed. Also called bus bar.

bus bar *n*: see *bus*.

bypass valve *n*: a valve that permits flow around a control valve, a piece of equipment, or a system.

C **cam** *n*: an eccentrically shaped disk, mounted on a camshaft, that varies in distance from its center to various points on its circumference. As the camshaft is rotated, a set amount of motion is imparted to a follower riding on the surface of the cam. In the internal-combustion engine, cams are used to operate the intake and exhaust valves.

camshaft *n*: the cylindrical bar used to support a rotating device called a cam.

carbon residue *n*: in a diesel fuel, the amount of carbon remaining in a special container after the fuel is burned under controlled conditions. Generally, high-quality fuels have low carbon residue.

centrifugal force *n*: the force that tends to pull all matter from the center of a rotating mass.

cetane number *n*: a measure of the ignition quality of fuel oil. The higher the cetane number, the more easily the fuel is ignited.

clearance *n*: 1. the distance by which one object clears another. 2. the amount of space between two objects.

combustion *n*: 1. the process of burning. Chemically, it is a process of rapid oxidation caused by the union of oxygen from the air with the material that is being oxidized or burned. 2. the organized and orderly burning of fuel inside the cylinder of an engine.

combustion cup *n*: on some diesel engines, a relatively small chamber, usually located just on top of the engine cylinder, into which fuel is injected. Since the cup also contains hot, compressed air, ignition occurs in the cup and the expanding gases push the piston down.

common bus *n*: typically, a large conductor made of copper cables that conducts alternating current from several alternators (AC generators) to rectifiers where the AC is converted to direct current (DC).

common rail *n*: the line in a certain type of fuel-injection system for a diesel engine that keeps fuel at a given pressure and feeds it through feed lines to each fuel injector.

common-rail injection *n*: a fuel-injection system on a diesel engine in which one line, or rail, holds fuel at a certain pressure and feed lines run from it to each fuel injector.

commutator *n*: a series of bars connected to the armature coils of an electric motor or generator. As the commutator rotates in contact with fixed brushes, the direction of flow of current to or from the armature is in one direction only.

compensator *n*: on a governor, a device that minimizes an engine's tendency to speed up and slow down as the governor seeks (hunts for) the correct engine speed. One type of compensator consists of a lever placed between the governor's speeder spring and a rod connected to a spring-loaded piston inside a cylinder, which is filled with oil. The lever places a large regulating force against the speeder spring to keep the spring and flyweights from moving rapidly and thus preventing small increases and decreases in engine speed.

compensator governor *n*: a type of engine governor that prevents hunting (an engine's speeding up and slowing down as it seeks to run at the speed dictated by the engine governor.) A compensator on the governor anticipates the engine's return to its set speed. When an engine's speed goes faster than the set speed, the compensator drops the engine's rpm; when engine speed drops below set speed, the compensator increases the engine's rpm. Normally, engine operators set the compensator to keep the drop small. With a small speed drop, the governor and compensator quickly make the engine go back to control speed. See *governor*, *hunting*.

compound *n*: a mechanism used to transmit power from the engines to the pump, the drawworks, and other machinery on a drilling rig. It is composed of clutches, chains and sprockets, belts and pulleys, and a number of shafts, both driven and driving. *v*: to connect two or more power-producing devices, such as engines, to run driven equipment, such as the drawworks.

compounded engines *n pl*: on mechanical drilling rigs, engines whose power is combined by means of machines, such as sprockets, chains, and belts, to drive equipment.

compressed air starting *n*: a method of starting a diesel engine in which an air compressor supplies high-pressure air to the engine's intake manifold. Since the air enters the engine under high pressure and temperature, combustion occurs when fuel is introduced and the engine starts.

compression ignition (CI) *n*: an ignition method used in diesel engines by which the air in the cylinder is compressed to such a degree by the piston that ignition occurs upon the injection of fuel. About a 1-pound (7-kilopascal) rise in pressure causes a 2°-F (1°-C) increase in temperature.

compression ratio *n*: 1. the ratio of the absolute discharge pressure from a compressor to the absolute intake pressure. 2. the ratio of the volume of an engine cylinder before compression to its volume after compression. For example, if a cylinder volume of 10 cubic inches (10 cubic centimetres) is compressed into 1 cubic inch (1 cubic centimetre), the compression ratio is 10:1.

compression stroke *n*: in a diesel engine, the upward movement of the piston in the cylinder in which the volume of air in the cylinder is reduced by the piston's moving upward in the cylinder. The piston compresses the air so much that it gets hot enough to ignite diesel fuel injected into the cylinder near the top of the piston's travel.

condensate trap *n*: a device installed on or near the fuel tank of a drilling rig engine into which the engine operator drains condensate water.

connecting rod *n*: 1. a forged-metal shaft that joins the piston of an engine to the crankshaft. 2. the metal shaft that is joined to the bull gear and crosshead of a mud pump.

contactor *n*: 1. a vessel or piece of equipment in which two or more substances are brought together. 2. a switch used to open or close an electric circuit.

control sleeve *n*: in a diesel engine's mechanical governor, a device that transfers the force of a spring and the centrifugal force of flyweights to a control lever that speeds up or slows down the engine. Usually, flyweight force tends to slow down the engine while spring force tends to speed up the engine.

control voltage *n*: in a diesel-electric system using alternators (AC generators), low powered electrical current that flows to a silicon controlled rectifier (SCR). When control voltage is applied to the SCR, the current generated by the AC generators flows through the SCR in one direction, thus converting alternating current to direct current. When control voltage is turned off, no current flows through the SCR.

coolant *n*: a cooling agent, usually a fluid, such as the liquid applied to the edge of a cutting tool to carry off frictional heat or a circulating fluid for cooling an engine.

coolant pump *n*: see *water pump.*

coolant temperature gauge *n*: a device on an engine, usually of the dial-and-needle type, that indicates the temperature of the engine's coolant at the point where the gauge's sensor is installed in the coolant system. Often, gauge sensors are installed to indicate coolant temperature at different places in the system; for example, where the coolant enters the engine's radiator or heat exchanger and where the coolant exits the engine after flowing through the cooling system.

corrosion *n*: any of a variety of complex chemical or electrochemical processes, e.g., rust, by which metal is destroyed through reaction with its environment.

corrosion inhibitor *n*: a chemical substance that minimizes or prevents corrosion in metal equipment.

crankcase *n*: the housing that encloses the crankshaft of an engine.

crankshaft *n*: a rotating shaft to which connecting rods are attached. It changes up and down (reciprocating) motion to circular (rotary) motion.

crank throw *n*: on a crankshaft, the highly polished and accurately machined portion of the crankshaft to which a piston rod is attached.

cutoff valve *n*: a special valve on an engine that, when activated, blocks the flow of fuel to the engine to make it stop running.

cycle *n*: the number of strokes a piston makes from one intake stroke to another intake stroke. Diesel engines may have either two strokes or four strokes per cycle.

D **demand** *n*: the quantity of oil, gas, or other petroleum products, or commodities (such as electricity) wanted at a specified time and price.

detergent *n*: in lubricating oils and in some engine fuels, a chemical that is added to the oil or to the fuel that suspends dirt, carbon, and other foreign matter in the oil or fuel. As a result of the detergents in motor oil, the oil will very quickly appear dirty because it is suspending the particles.

diesel engine *n*: a high-compression, internal-combustion engine used extensively for powering drilling rigs. In a diesel engine, air is drawn into the cylinders and compressed to very high pressures; ignition occurs as fuel is injected into the compressed and heated air. Combustion takes place within the cylinder above the piston, and expansion of the combustion products imparts power to the piston.

diode *n*: 1. an electronic device that restricts current flow chiefly in one direction. 2. a radio tube that contains an anode and a cathode.

direct-acting electric actuator *n*: a device on a diesel engine's governor that increases the engine's speed by increasing positive voltage going to the actuator and governor. As the engine needs more fuel to go faster, the direct-acting actuator increases positive voltage to make the governor increase the fuel flowing to the fuel injectors. See *electrically-actuated governor*.

direct current (DC) *n*: electric current that flows in only one direction. Compare *alternating current*.

dissociation *n*: the separation of a molecule into two or more fragments (atoms, ions) by interaction with another body or by the absorption of electromagnetic radiation.

distillation *n*: the process of driving off gas or vapor from liquids or solids, usually by heating, and condensing the vapor back to liquid to purify, fractionate, or form new products.

distributor *n*: a device that directs the proper flow of fuel or electrical current to the proper place at the proper time in the proper amount.

dry air cleaner *n*: on an engine, a device that contains an air filter element that does not depend on oil to effectively filter the air entering the engine. Instead, the filter element has a number of folds and chambers that trap dust and dirt going into the engine's air intake. Some dry elements may be cleaned and reinstalled; others are discarded and replaced. Compare *wet air cleaner*.

E

electrically-actuated governor *n*: a hydraulic governor on an engine that has a reversible electric motor (it runs both clockwise and counterclockwise). By manipulating a remote control, the engine operator can adjust the electric motor to closely control the engine's speed. See *governor, hydraulic governor*.

electric starter *n*: a device that uses a battery, an electric motor, gears, and cables to provide a way of starting an engine. An electrically actuated motor turns the engine over by means of a pinion gear in the starter that engages a ring gear on the engine flywheel.

electrolyte *n*: 1. a chemical that, when dissolved in water, dissociates into positive and negative ions, thus increasing its electrical conductivity. See *dissociation*. 2. the electrically conductive solution that must be present for a corrosion cell to exist.

engine *n*: a machine for converting the heat content of fuel into rotary motion that can be used to power other machines. Compare *motor*.

engine temperature switch (ETS) *n*: a device on an engine that senses overheating and shuts down the engine if overheating occurs.

exchanger *n*: a piping arrangement that permits heat from one fluid to be transferred to another fluid as they travel countercurrently to one another. In the heat exchanger of an emulsion-treating unit, heat from the outgoing clean oil is transferred to the incoming well fluid, cooling the oil and heating the well fluid. In the heat exchanger of a glycol dehydration unit, heat from the hot lean glycol flows through the inner flow tube in the opposite direction of the cool rich glycol, which flows through a shell built around the tube.

exciter *n*: a small DC generator mounted on top of a main generator to produce the field for the main generator.

exhaust *n*: the burned gases that are removed from the cylinder of an engine. *v*: to remove the burned gases from the cylinder of an engine.

exhaust manifold *n*: a piping arrangement, immediately adjacent to the engine, that collects burned gases from the engine and channels them to the exhaust pipe.

exhaust pipe *n*: on an engine, flexible steel tubing that connects the engine exhaust manifold outlet to the muffler. See *muffler*.

exhaust silencer *n*: see *muffler*.

exhaust stack *n*: see *tail pipe*.

exhaust stroke *n*: in an engine, the movement of the piston during which time it pushes burned fuel gases out of the cylinder.

exhaust system *n*: valves (or ports), manifolds (passageways and connections), piping, noise silencers (mufflers), and other devices, such as exhaust stacks, that serve to remove burned fuel gases from an engine.

exhaust valve *n*: the cam-operated mechanism through which burned gases are ejected from an engine cylinder.

explosion cover *n*: see *explosion door*.

explosion door *n*: on an engine, one of usually several spring-loaded, lightweight metal plates placed over openings in the engine's crankcase. Oxygen in the air entering the base of an engine mixes with the oil there. A hot spot could cause the oil and oxygen to explode and damage the crankcase. To prevent such damage, the spring-operated explosion covers (doors) open to release the pressure from the explosion. They then slam closed to prevent more air from entering. Also called explosion cover.

F **filter** *n*: a porous medium through which a fluid is passed to separate particles of suspended solids from it.

fin *n*: a thin, sharp ridge around the box or the pin shoulder of a tool joint, caused by the use of boxes and pins with different-sized shoulders. See *radiator fin*.

fire point *n*: the temperature at which a petroleum product burns continuously after being ignited. See *flash point*.

firing order *n*: the sequence in which combustion occurs in a multicylinder engine. For example, in an eight cylinder engine, the firing order could be 1–8–4–3–6–5–7–2, which means that combustion occurs first in the first cylinder, then in the eighth cylinder, and so on, until combustion occurs in the second cylinder; then, the sequence starts over.

flash point *n*: the temperature at which a petroleum product ignites momentarily but does not burn continuously. Compare *fire point*.

flux *n*: the lines in a magnetic field.

flyweight *n*: on a mechanical engine governor, one of usually two small metal weights that spin as the engine runs. When the engine speeds up, centrifugal force on the spinning flyweights increases, which causes a spring to compress and slow the engine down. Conversely, when the engine slows down, centrifugal force on the flyweights decreases which causes the spring to expand and speed the engine up.

flywheel *n*: a large, circular disk, connected to and revolving with an engine crankshaft. It stores energy and disburses it as the engine runs.

four-stroke/cycle engine *n*: an engine in which the piston moves from top dead center to bottom dead center two times to complete a cycle of events. The crankshaft must make two complete revolutions, or 720°.

fuel centrifuge *n*: a device an engine operator uses to separate water and solid materials from fuel. Centrifugal force created by the rapidly spinning centrifuge causes dirt and water, which are heavier (denser) than fuel, to move to the outside of the centrifuge where they are removed.

fuel-injection nozzle *n*: see *nozzle*.

fuel injector *n*: a mechanical device that sprays fuel into a cylinder of an engine at the end of the compression stroke.

fuel knock *n*: a hammerlike noise produced when fuel is not burned properly in a cylinder.

fuel modulator *n*: a device installed on a diesel engine to reduce the amount of smoke coming out of the engine's exhaust. If the engine's governor delivers more fuel than air to the engine, the engine smokes too much. A fuel modulator makes the governor increase the fuel supply only at the same rate as the air increase. Such rate control holds down the black smoke from the engine exhaust during acceleration or sudden loading. See *governor*.

fuel pump *n*: the pump that pressurizes fuel to the pressure used for injection. In a diesel engine the term is used to identify several different pumps: it is loosely used to describe the pump that transfers fuel from the main storage tank to the day tank; it is also used to describe the pump that supplies pressure to the fuel-injection pumps, although this is actually a booster-type pump.

fuel transfer pump *n*: any relatively small pump in an engine's fuel system that moves fuel from one fuel tank to another or from a tank to another location in the fuel system.

G

gasket *n*: any material (such as paper, cork, asbestos, or rubber) used to seal two essentially stationary surfaces.

gasoline engine starter motor *n*: on a diesel engine, a relatively small engine that runs on spark plugs and gasoline and whose power is used to turn over (move) the pistons in the diesel. As the diesel's pistons move, pressure and heat builds in the diesel's cylinders and the diesel starts. As the diesel begins running, the gasoline engine disengages from the diesel.

gate voltage *n*: in an AC-DC diesel-electric system, the small amount of electricity applied to a silicon controlled rectifier (SCR) to cause electricity to flow in one direction through the SCR. As long as gate voltage is applied, the SCR converts AC voltage to DC voltage as electricity flows through the SCR. If gate voltage is shut off, no electrical current can flow through the SCR.

generator *n*: a machine that changes mechanical energy into electrical energy.

glow plug *n*: a small electric heating element placed inside a diesel engine cylinder to heat the air and to make starting easier.

GM Hydrostarter TM *n*: a hydraulic starter motor manufactured by General Motors (GM) that uses hydraulic fluid (light oil) to power a relatively small motor attached to the diesel engine. To start the diesel, the operator activates the Hydrostarter which turns over (moves) the diesel's pistons. As the pistons move, pressure and heat builds up in the diesel's cylinders and the diesel starts.

governor *n*: any device that limits or controls the speed of an engine.

H

header *n*: a chamber from which fluid is distributed to smaller pipes or conduits, e.g., a manifold.

heat exchanger *n*: see *exchanger*.

horsepower (hp) *n*: a unit of measure of work done by a machine. One horsepower equals 33,000 foot-pounds per minute. (Kilowatts are used to measure power in the international, or SI, system of measurement.)

hp *abbr*: horsepower.

hunting *n*: a surge of engine speed to a higher number of revolutions per minute (rpm), followed by a drop to normal speed without manual movement of the throttle. It is often caused by a faulty or improperly adjusted governor.

hydraulic *adj*: 1. of or relating to water or other liquid in motion. 2. operated, moved, or effected by water or liquid.

hydraulic governor *n*: a governor on an engine that operates by means of oil inside its housing. Unlike a mechanical governor, which is mechanically linked to the engine's speed control, a hydraulic governor operates the speed control with oil pressure inside the governor. See *governor*. Compare *mechanical governor*.

hydraulic starter *n*: on an engine, a device used to start the engine that uses hydraulic fluid under pressure to operate a motor on the starter. When engaged, the starter motor turns the engine's flywheel to make the engine start.

hydrocarbons *n pl*: organic compounds of hydrogen and carbon whose densities, boiling points, and freezing points increase as their molecular weights increase. Although composed of only two elements, hydrocarbons exist in a variety of compounds, because of the strong affinity of the carbon atom for other atoms and for itself. The smallest molecules of hydrocarbons are gaseous; the largest are solids. Petroleum is a mixture of many different hydrocarbons.

hydrometer *n*: an instrument with a graduated stem, used to determine the gravity of liquids. The liquid to be measured is placed in a cylinder, and the hydrometer dropped into it. It floats at a certain level in the liquid (high if the liquid is light, low if it is heavy), and the stem markings indicate the gravity of the liquid.

I

idle *v*: to operate an engine without applying a load to it.

ignition quality *n*: the ability of a fuel to ignite when it is injected into the compressed-air charge in a diesel engine cylinder. It is measured by an index called the cetane number.

immersion heater temperature switch (IHTS) *n*: a temperature-sensitive device installed on engines in extremely cold climates, such as in Alaska and Siberia. Before the engine is started, the switch senses very cold coolant and turns on an immersion heater in the coolant tank, thus ensuring that the engine warms up quickly.

impeller *n*: a set of mounted blades used to impart motion to a fluid (e.g., the rotor of a centrifugal pump).

injection *n*: the process of forcing fluid into something. In a diesel engine, the introduction of high-pressure fuel oil into the cylinders.

injection line *n*: strong steel tubing that conducts fuel from the fuel tanks to fuel injectors on the engine.

injector *n*: a fuel atomizing device that injects (puts) a fine spray of fuel into the combustion chamber of an engine.

injector pump *n*: a chemical feed pump that injects chemical reagents into a flow-line system to treat emulsions at a rate proportional to that of the flow of the well fluid. Operating power may come from electric motors or from linkage with the walking beam of a pumping well.

intake stroke *n*: the downward movement of a piston in a cylinder that creates an area of low pressure inside the cylinder. The low pressure draws in air from the atmosphere (or from a blower).

intake valve *n*: 1. the cam-operated mechanism on an engine through which air and sometimes fuel are admitted to the cylinder. 2. on a mud pump, the valve that opens to allow mud to be drawn into the pump by the pistons moving in the liners.

isochronous governor *n*: a governor that maintains a constant speed of the prime mover regardless of the load applied, within the capacity of the prime mover.

K

kick *n*: an entry of water, gas, oil, or other formation fluid into the wellbore during drilling. It occurs because the pressure exerted by the column of drilling fluid is not great enough to overcome the pressure exerted by the fluids in the formation drilled. If prompt action is not taken to control the kick, or kill the well, a blowout may occur. See *blowout*.

kilowatt *n*: a metric unit of power equal to approximately 1.34 horsepower.

L

lead-acid battery *n*: a storage battery in which the electrodes are grids of lead oxides that change in composition during charging and discharging, and the electrolyte is dilute sulfuric acid.

lifter *n*: a device in an engine against which a cam rotates as the engine runs. As the high point of the cam rotates, it pushes against (lifts) the lifter. The lifter, in turn, actuates a push rod or other device to open a valve or similar device. Sometimes called a cam follower.

load *n*: 1. in mechanics, the weight or pressure placed on an object. The load on a bit refers to the amount of weight of the drill collars allowed to rest on the bit. See *weight on bit*. 2. in reference to engines, the amount of work that an engine is doing; for example, 50 percent load means that the engine is putting out 50 percent of the power that it is able to produce. 3. the amount of gas delivered or required at any specified point or points on a system; load originates primarily at the gas-consuming equipment of the customer. *v*: 1. to engage an engine so that it works. Compare *idle*. 2. to set a governor to maintain a given pressure as the rate of gas flow through the governor varies. Compare *demand*.

lubricant *n*: a substance—usually petroleum-based—that is used to reduce friction between two moving parts.

lugging power *n*: the torque, or turning power, delivered to the flywheel of a diesel engine.

M

magnetic flux *n*: energy that flows between the north and south poles of a magnet. The magnet may be either natural or it may be an electromagnet, which is created by applying electric current to a conductor, such as iron or steel. See *flux*.

main bearing *n*: in an engine, a large circular friction-reducing device installed on the engine's crankshaft. Main bearings are mounted in the engine's crankcase and the crankshaft rotates on them. Engine oil lubricates them as the engine runs. Most main bearings are plain bearings, in that their wear surface is flat or plain and do not have balls or rollers. Compare *ball bearing, roller bearing*.

manifold *n*: 1. an accessory system of piping to a main piping system (or another conductor) that serves to divide a flow into several parts, to combine several flows into one, or to reroute a flow to any one of several possible destinations. 2. a pipe fitting with several side outlets to connect it with other pipes. 3. a fitting on an internal-combustion engine made to receive exhaust gases from several cylinders.

manifold pressure gauge *n*: a device installed on an engine's intake manifold that indicates the pressure inside the manifold. Manifold pressure is a measure of the airflow into the engine cylinders at a given speed and intake air temperature. Thus, when combined with the engine's rpm, manifold pressure is an indication of the engine's power output.

manometer *n*: a U-shaped piece of glass tubing containing a liquid (usually water or mercury) that is used to measure the pressure of gases or liquids. When pressure is applied, the liquid level in one arm rises while the level in the other drops. A set of calibrated markings beside one of the arms permits a pressure reading to be taken, usually in inches or millimetres.

mechanical governor *n*: a speed-control device on an engine. Mechanical governors consist of flyweights, springs, and mechanical connections to the engine's speed control. Compare *electrically-actuated governor, hydraulic governor*. See *governor*.

metal-edge strainer *n*: in an engine, a fuel filter that consists of several very thin metal discs stacked inside a housing. As fuel flows through the strainer, foreign matter in the fuel is trapped by the very small spaces between the discs.

micropore paper *n*: heavy-duty paper perforated with several very small holes (pores). Folded in accordion pleats, micropore paper often serves as a secondary filter element in an engine's fuel system. Fuel passes through the pores, while the unperforated part of the paper stops dirt.

motor *n*: a hydraulic, air, or electric device used to do work. Compare *engine*.

muffler *n*: a device installed on an engine to quiet the barking sound produced by exhaust gases exiting through the exhaust pipe of the engine. One type of muffler is a steel cylinder with baffle plates. The baffle plates, flat steel sheets welded inside the cylindrical body of the muffler, change the direction of exhaust gas flow. Changing the direction of flow allows the gases to expand gradually, rather than all at once. Gradual expansion is quieter than rapid expansion. Sometimes called an exhaust silencer.

multipump injection system *n*: in a diesel engine, a system that uses several fuel pumps to take fuel from a supply tank and send it to the engine's fuel injectors. The pumps may be separate or combined into a single housing.

naturally aspirated *adj*: term used to describe an internal combustion engine in which air flows into the engine by means of atmospheric pressure only. Compare *supercharge*.

naturally aspirated engine *n*: an engine in which the air enters the engine's cylinders by the simple action of the pistons moving downward in the cylinders. Compare *supercharged engine*.

needle valve *n*: a globe valve that contains a sharp-pointed, needlelike plug that is driven into and out of a cone-shaped seat to control accurately a relatively small rate of fluid flow. In a fuel injector, the fuel pressure forces the needle valve off its seat to allow injection.

nozzle *n*: 1. a passageway through jet bits that causes the drilling fluid to be ejected from the bit at high velocity. The jets of mud clear the bottom of the hole. Nozzles come in different sizes that can be interchanged on the bit to adjust the velocity with which the mud exits the bit. 2. the part of the fuel system of an engine that has small holes in it to permit fuel to enter the cylinder. Properly known as a fuel-injection nozzle, but also called a spray valve. The needle valve is directly above the nozzle.

oil-bath air cleaner *n*: on an engine, a canister that has a relatively small amount of lubricating oil in the bottom and which filters air entering an engine. The running engine draws air through the cleaner, where the air passes through an element and over the oil bath. Dust and dirt particles in the intake air are removed by the element and by the oil bath.

oil-bath cleaner *n*: see *oil-bath air cleaner*.

oil cooler *n*: on an engine, a device through which engine lubricating oil is circulated to reduce its temperature to acceptable levels. Some oil coolers depend on the surrounding air to reduce the temperature, but on most large engines, the oil is circulated through tubes that are surrounded by engine coolant.

oil pressure gauge *n*: a device that shows the pressure of the oil in an engine's lubricating system. Oil must achieve a certain minimum pressure to move throughout the engine and to provide an adequate film of oil around the engine's moving parts.

oil pump *n*: a special pump, usually of the gear type, that moves oil through an engine. The pump's intermeshing gears rotate to build pressure and circulate the oil.

oil temperature gauge *n*: a device that shows the temperature of the oil in an engine's lubricating system. Oil whose temperature is too high cannot lubricate properly.

orifice *n*: an opening of a measured diameter that is used for measuring the flow of fluid through a pipe or for delivering a given amount of fluid through a fuel nozzle. In measuring the flow of fluid through a pipe, the orifice must be of smaller diameter than the pipe diameter. It is drilled into an orifice plate held by an orifice fitting.

oscilloscope *n*: a test instrument that visually records an electrical wave on a fluorescent screen.

overspeed governor *n*: a special type of engine governor that prevents an engine from running too fast (overspeeding) by shutting down the engine if it overspeeds. See *governor*.

overspeeding *n*: an engine's running beyond the maximum speed for which it was designed. Overspeeding an engine can severely damage it.

overspeed trip *n*: see *overspeed governor*.

overspeed trip device *n*: a special type of engine governor that prevents an engine from running too fast and possibly destroying itself if the regular governor fails.

P

paraffin *n*: a saturated aliphatic hydrocarbon having the formula C_nH_{2n+2} (e.g., methane, CH_4; ethane, C_2H_6). Heavier paraffin hydrocarbons (i.e., $C_{18}H_{38}$) form a waxlike substance that is called paraffin. These heavier paraffins often accumulate on the walls of tubing and other production equipment, restricting or stopping the flow of the desirable lighter paraffins.

pigtail *n*: in the brush holder of a generator, a device composed of several braided copper wires that conduct electricity from a generator's brush to the generator's wiring.

pinion *n*: 1. a gear with a small number of teeth designed to mesh with a larger wheel or rack. 2. the smaller of a pair or the smallest of a train of gear wheels.

pipe rack *n*: a horizontal support for tubular goods.

piston *n*: a cylindrical sliding piece that is moved by or that moves against fluid pressure within a confining cylindrical vessel.

piston stroke *n*: the length of movement, in inches (millimetres), of the piston of an engine from top dead center (TDC) to bottom dead center (BDC).

plunger *n*: 1. a basic component of the sucker rod pump that serves to draw well fluids into the pump. 2. the rod that serves as a piston in a reciprocating pump. 3. the device in a fuel-injection unit that regulates the amount of fuel pumped on each stroke.

positive crankcase ventilation (PCV) valve *n*: a valve installed on an engine that, when open, directs gases from inside the engine through piping to the intake manifold. At the intake manifold, the crankcase gases enter the engine's combustion chamber and are burned. Burning crankcase gases in this manner not only cuts down on air pollution, but also scavenges corrosive fumes from the engine to prevent sludge from forming.

pour point *n*: the lowest temperature at which a fuel will flow. For oil, the pour point is a temperature $5°F$ ($-15°C$) above the temperature at which the oil is solid.

pressure gauge *n*: an instrument that measures fluid pressure and usually registers the difference between atmospheric pressure and the pressure of the fluid by indicating the effect of such pressures on a measuring element (e.g., a column of liquid, pressure in a Bourdon tube, a weighted piston, or a diaphragm).

pressure relief valve *n*: a valve that opens at a preset pressure to relieve excessive pressures within a vessel or line. Also called a pop valve, relief valve, safety valve, or safety relief valve.

pressurized cooling system *n*: an engine cooling system in which a pressure-tight seal is maintained, usually by a special cap placed on the radiator's opening. The pressure-tight seal keeps the pressure on the cooling system slightly above atmospheric pressure. Maintaining pressure slightly above that of the atmosphere raises the boiling point of water (often the main ingredient in the engine's coolant), eliminates evaporation of coolant from the system, and permits a higher operating coolant temperature, which results in more effective heat transfer from the coolant to the air.

prestart lubrication system *n*: an assembly of devices, including a special oil pump that works separately from the engine, that allow an engine operator to circulate lubricating oil through an engine prior to starting it. Pressuring up the oil in an engine before it starts ensures that a good lubricating oil film forms on the engine parts, thus reducing wear on them.

primary pump *n*: see *booster pump*.

pushrod *n*: a device used to link the valve to the cam in an engine.

pushrod guide *n*: in an engine's valve train, a hollow cylindrical opening or tube through which valve pushrods move up and down. In many engines, the cam on the engine camshaft moves a pushrod that contacts a rocker arm. The rocker arm, in turn, contacts the valve stem, which opens the valve.

pyrometer *n*: an instrument for measuring temperatures, especially those above the range of mercury thermometers.

R

rack *n*: 1. framework for supporting or containing a number of loose objects, such as pipe. See *pipe rack*. 2. a bar with teeth on one face for gearing with a pinion or worm gear. 3. a notched bar used as a ratchet. *v*: 1. to place on a rack. 2. to use as a rack.

rack-and-pinion gear *n*: a gear comprising a bar with teeth on one face that engages with a pinion (a small gear).

radiator *n*: an arrangement of pipes that contains a circulating fluid and is used for heating an external object or cooling an internal substance by radiation.

radiator core *n*: on an engine, the tube or tubes through which engine coolant is circulated. These tubes bend back and forth several times so that the coolant will have plenty of surface to contact as it flows through them. Very thin metal plates (fins) are attached to the tube, which radiate heat in the coolant to the surrounding air.

radiator fin *n*: on an engine, very thin metal plates that are attached to the radiator's tubes. Because there are hundreds of these plates in contact with the surrounding air, they efficiently radiate heat from the coolant circulating through the tubes.

ram blowout preventer *n*: a blowout preventer that uses rams to seal off pressure on a hole that is with or without pipe. Also called a ram preventer. Compare *annular blowout preventer*.

raw water *n*: in an engine's heat exchanger, water that circulates outside and around the tubes through which the engine's coolant circulates. Raw water is untreated water that removes heat from the engine's coolant. Offshore, raw water is often seawater. Since raw water does not contact engine parts, it usually does not require treatment.

reciprocating pump *n*: a pump consisting of a piston that moves back and forth or up and down in a cylinder. The cylinder is equipped with inlet (suction) and outlet (discharge) valves. On the intake stroke, the suction valves are opened, and fluid is drawn into the cylinder. On the discharge stroke, the suction valves close, the discharge valves open, and fluid is forced out of the cylinder.

reciprocation *n*: a back-and-forth or up-and-down movement (as the movement of a piston in an engine or pump).

rectifier *n*: a device used to convert alternating current into direct current.

rectify *v*: to change an alternating current to a direct current.

relative density *n*: 1. the ratio of the weight of a given volume of a substance at a given temperature to the weight of an equal volume of a standard substance at the same temperature. For example, if 1 cubic inch of water at 39°F (3.9°C) weighs 1 unit and 1 cubic inch of another solid or liquid at 39°F weighs 0.95 unit, then the relative density of the substance is 0.95. In determining the relative density of gases, the comparison is made with the standard of air or hydrogen. 2. the ratio of the mass of a given volume of a substance to the mass of a like volume of a standard substance, such as water or air.

relief valve *n*: see *pressure relief valve*.

reverse-acting electric actuator *n*: a device that increases an engine's speed by decreasing the positive voltage going to the electric actuator and governor. As the engine needs more fuel to go faster, the reverse-acting actuator decreases the positive voltage to make the governor increase the fuel. Compare *direct-acting electric actuator*. See *governor*.

reverse current relay *n*: a device installed in a diesel-electric system that disconnects a generator from the system when an engine is underloaded and the generator attached to it behaves like a motor. A generator acting like a motor draws current from the system, instead of supplying current. The generator acting like a motor then drives the underloaded engine, which can severely damage the engine.

rheostat *n*: a resistor that is used to vary the electrical current flow in a system.

rocker arm *n*: a bell-crank device that transmits the movement of the pushrod to the valves in an engine.

roller bearing *n*: a bearing in which a finely machined shaft (the journal) rotates in contact with a number of cylinders (rollers). Compare *ball bearing*.

Roots blower *n*: a special compressor used on two-stroke engines to supercharge the engine's intake air and scavenge (remove) exhaust gases from the engine cylinders. The blower has two corkscrew-shaped (helical) rotors with blades (lobes) that rotate inside a housing. The engine drives the rotors with gears at about twice the engine's speed. The rotors rotate in opposite directions at the same speed. As they rotate, the lobes compress air drawn through an air cleaner on top of the housing. The compressed air exits the blower from the bottom or the side of the housing and goes into the engine's air intake manifold.

rotor *n*: 1. a device with vanelike blades attached to a shaft. The device turns or rotates when the vanes are struck by a fluid directed there by a stator. 2. the rotating part of an induction-type alternating current electric motor.

Saybolt Second Universal (SSU) *n*: a unit for measuring the viscosity of lighter petroleum products and lubricating oils. See *Saybolt viscometer*.

Saybolt viscometer *n*: an instrument used to measure the viscosity of fluids, consisting basically of a container with a hole or jet of a standard size in the bottom. The time required for the flow of a specific volume of fluid is recorded in seconds at three temperatures (100°F–37.8°C, 130°F–54.4°C, and 210°F–98.9°C). The time measurement unit is referred to as the Saybolt Second Universal (SSU).

scavenge *v*: to remove exhaust gases from a cylinder by forcing compressed air in and exhaust gases out. Such removal takes place in all two-cycle diesel engines.

scavenging efficiency *n*: in two-strokes-per-cycle engines, a measure of the engine blower's ability to remove exhaust gases from the engine cylinder after combustion.

sensitivity *n*: a measure of an engine governor's ability to sense a change in an engine's speed and make an adjustment to correct the speed change.

shell *n*: 1. the body of a tank. 2. the horizontal tank on a tank car that contains the liquid being transported. 3. the steel backing of a precision insert bearing on a bit.

shim *n*: a thin, often tapered piece of material used to fill in space between things (such as for support, leveling, or adjustment of fit).

shutdown *n*: the act of stopping a machine or device from running. For example, engine operators perform a shutdown when they stop an engine.

silicon-controlled rectifier (SCR) *n*: a device that changes alternating current to direct current by means of a silicon control gate.

sine wave *n*: a graphic mathematical representation of the wave form of alternating current or voltage.

slip ring *n*: a conducting ring that gives current to or receives current from the brushes in a generator or motor.

solenoid *n*: a cylindrical coil of wire that resembles a bar magnet when it carries a current so that it draws a movable core into the coil when the current flows.

spark ignition (SI) *n*: ignition of a fuel-air mixture by means of a spark discharged by a spark plug.

spark plug *n*: a device that fits into the cylinder of an internal-combustion engine and that provides the spark for ignition of the fuel-air mixture during the combustion stroke of the piston. It carries two electrodes separated by an air gap; current from the ignition system discharges across the gap to form the spark.

specific gravity *n*: see *relative density*.

speed droop *n*: the number of revolutions per minute that an engine slows down from running at maximum no-load speed to running at maximum full-load speed. Speed droop should not exceed 7%.

speeder spring *n*: a small spring inside an engine governor that counteracts the force of flyweights in the governor. The speeder spring moves down on a sleeve in the governor to increase the fuel supply to the engine; at the same time, the flyweights move the sleeve up to decrease the fuel supply. Since the flyweights spin at a speed determined by the engine's speed, if engine speed drops, the speeder spring moves down to speed the engine up. See *flyweight, governor*.

speed limiter *n*: a type of engine governor that prevents an engine from exceeding a set speed. Usually, limiters simply prevent the engine from running too fast to prevent damage to the engine; they do not shut down the engine.

spring-loaded centrifugal governor *n*: a mechanical governor that includes a special spring, called a speeder spring, that offsets the centrifugal force of spinning flyweights in the governor. In many governors, the flyweights tend to make the governor slow the engine down, while the spring tends to make the governor speed the engine up. Centrifugal force and spring pressure balance each other to maintain the engine's speed.

spring shunt *n*: on a generator's brush holder, a braided copper wire conductor that conducts (shunts) any current that may build up the brush holder's spring away from the spring.

stack *n*: 1. a vertical arrangement of blowout prevention equipment. Also called preventer stack. See *blowout preventer*. 2. the vertical chimney-like installation that is the waste disposal system for unwanted vapor such as flue gases or tail-gas streams. See *exhaust stack*.

starter *n*: on an engine, an electrical, hydraulic, air, or other device used to rotate the engine's flywheel or pistons so that the fuel, air, and (in some cases) spark can enter the engine and make it begin running on its own.

stator *n*: 1. a device with vanelike blades that serves to direct a flow of fluid (such as drilling mud) onto another set of blades (called the rotor). The stator does not move; rather, it serves merely to guide the flow of fluid at a suitable angle to the rotor blades. 2. the stationary part of an induction-type alternating-current electric motor. Compare *rotor*.

storage battery *n*: a series of storage cells that produce electricity by chemical action of acid or alkaline solution on metallic plates. Charging the battery with DC electricity in the opposite direction restores the chemical condition necessary for further output of electricity.

strainer *n*: 1. a part of a LACT unit that removes foreign particles from the crude, which might disrupt the operation of close-tolerance moving parts. 2. a device placed upstream of a meter to remove foreign material from the stream that might damage the meter or interfere with its operation. The strainer element is generally coarser than a filter designed to remove solid contaminants.

stroke *n*: the up and down (reciprocating) movement of a piston in a cylinder.

strokes-per-cycle *n pl*: the number of strokes an engine piston makes inside a cylinder to complete one firing cycle. Most engines are either two-strokes-per-cycle or four-strokes-per-cycle. In a two-strokes-per-cycle engine, the engine crankshaft makes two revolutions to complete one cycle. As the crankshaft moves the piston down on the first stroke, fuel is injected and combustion and power occur. As the crankshaft moves the piston up on the second stroke, burned gases go to exhaust, air is forced into the cylinder, and the piston compresses the air as it moves up the cylinder. In a four-strokes-per-cycle engine, the crankshaft makes four revolutions to complete one cycle. As the crankshaft moves the piston down on the first stroke, the piston draws in air (or it is forced in with a blower). As the crankshaft moves the piston up, the piston compresses the air in the cylinder; just before the piston reaches the top of its travel, fuel is injected. Combustion of the fuel-air mixture creates power and moves the piston down. Finally, as the piston moves up on the fourth stroke, the piston pushes the burned gases into the exhaust system.

supercharge *v*: to supply a charge of air to the intake of an internal-combustion engine at a pressure higher than that of the surrounding atmosphere.

supercharged engine *n*: an engine in which a compressor raises the pressure of the air and forces it into the engine's cylinders.

supercharger *n*: a device that compresses atmospheric air and forces it into an engine. Raising the pressure of an engine's intake air makes it denser and thus delivers more oxygen into the cylinder. More oxygen produces more power in the combustion process.

synchronous speed *n*: in an electric motor, the speed of the rotating field:

$$\text{synchronous speed (rpm)} = \frac{120 \times \text{frequency (hertz)}}{\text{number of poles}}$$

synchroscope *n*: a device used by an engine operator to ensure that all the engines on a rig are running at synchronous speed. Ensuring synchronous speed is necessary to avoid damaging a generator when taking it off line or putting it on line.

T

tachometer (tach) *n*: an instrument that measures the speed of rotation of an engine.

tail pipe *n*: 1. a pipe run in a well below a packer. 2. a pipe used to exhaust gases from the muffler of an engine to the outside atmosphere.

temperature regulator *n*: see *thermostat*.

thermostat *n*: a control device used to regulate temperature.

Thyristor *n*: see *silicon-controlled rectifier*.

timing *n*: the relationship of all moving parts in an engine. Each part depends on another, so all parts must operate in the right relation with each other as the engine turns.

timing gear train *n*: a set of gears in an engine, driven by the crankshaft, set to drive the equipment necessary for the engine's operation. Engine equipment driven by the timing gears includes the oil pump, intake and exhaust valve mechanisms, fuel injectors, and magnetos.

top dead center (TDC) *n*: the position of a piston when it is at the highest point possible in the cylinder of an engine; often marked on the flywheel.

torque *n*: the turning force that is applied to a shaft or other rotary mechanism to cause it to rotate or tend to do so. Torque is measured in foot-pounds, joules, newton-metres, and so forth.

total volume *n*: in diesel engine fuel tanks, the full capacity of the tank; usually, not all the fuel in the tank is usable. Compare *useful volume*.

turbine *n*: a bladed rotor flowmeter component that turns at a speed that is proportional to the mean velocity of the stream and therefore to the volume rate of flow.

turbocharger *n*: a centrifugal blower driven by exhaust gas turbines and used to supercharge an engine.

two-stroke diesel engine *n*: an engine in which the piston moves from top dead center to bottom dead center and then back to top dead center to complete a cycle. Thus, the crankshaft must turn one revolution, or 360°.

U

union *n*: a coupling device that allows pipes to be connected without being rotated. The mating surfaces are pulled together by a flanged, threaded collar on the union.

unit fuel-injector assembly *n*: a device that combines an internal high-pressure fuel pump, an injector valve, and a spray nozzle, and that is used to inject diesel fuel into an engine's combustion chamber.

useful volume *n*: in diesel engine fuel tanks, that part of the tank's capacity from which the fuel system can take fuel. Useful volume is less than the actual or total volume of the tank. Compare *total volume*.

V

vacuum *n*: 1. a space that is theoretically devoid of all matter and that exerts zero pressure. 2. a condition that exists in a system when pressure is reduced below atmospheric pressure.

valve overlap *n*: the phenomenon that occurs during the operating cycle of an engine in which both the intake and exhaust valves are open at the same time. Valve overlap occurs in a four-strokes-per-cycle engine at the end of the engine's exhaust stroke and the beginning of the intake stroke. Both valves being open allow exhaust gases to finish leaving the cylinder and, at the same time, allow inlet air to begin filling the cylinder. Valve overlap ensures that the cylinder is completely filled with fresh air; also, the cool incoming air helps cool the hot exhaust valve.

viscosity *n*: a measure of the resistance of a fluid to flow. Resistance is brought about by the internal friction resulting from the combined effects of cohesion and adhesion. The viscosity of petroleum products is commonly expressed in terms of the time required for a specific volume of the liquid to flow through a capillary tube of a specific size at a given temperature.

volatility *n*: the tendency of a liquid to assume the gaseous state.

voltage *n*: potential difference or electromotive force, measured in volts.

voltmeter *n*: an instrument used to measure, in volts, the difference of potential in an electrical circuit.

water jacket *n*: in the cooling system, a passageway inside the rim for circulating water.

W

water pump *n*: on an engine, a device, powered by the engine, that moves coolant (water) through openings in the engine crankcase, through the radiator or heat exchanger, and back into the crankcase.

weight on bit (WOB) *n*: the amount of downward force placed on the bit by the weight of the drill collars.

wet air cleaner *n*: see *oil-bath air cleaner*.

winding *n*: a set of conductors installed to form the current-carrying element of a dynamo or a stationary transformer.

worm gear *n*: the gear of a worm (a short revolving screw with spiral-shaped threads) and a worm wheel (a toothed wheel gearing with the thread of a worm) working together.

wrist pin *n*: in an engine, the hard steel, hollow cylinder that attaches the piston to the piston rod. A circular opening in the piston is lined up with a corresponding circular opening in the rod and the wrist pin is pushed through the openings. Usually, special keys on each end of the pin lock the pin in the piston. Also called piston pin.